Precautionary Labels for Chemical Containers

Neal Langerman, Ph.D.
Chemical Safety Associates, Inc.
San Diego, California

CRC Press
Taylor & Francis Group
Boca Raton London New York

CRC Press is an imprint of the
Taylor & Francis Group, an **informa** business

First published 1994 by CRC Press
Taylor & Francis Group
6000 Broken Sound Parkway NW, Suite 300
Boca Raton, FL 33487-2742

Reissued 2018 by CRC Press

Library of Congress Cataloging-in-Publication Data

Langerman, Neal
Precautionary labels for chemical containers/Neal Langerman
p. cm.
Includes index.
ISBN 0-87371-917-4
1. Chemicals—Labeling. I. Title.
TP149.L27 1994
660. '2804—dc20 93-44587

A Library of Congress record exists under LC control number: 93044587

Publisher's Note
The publisher has gone to great lengths to ensure the quality of this reprint but points out that some imperfections in the original copies may be apparent.

Disclaimer
The publisher has made every effort to trace copyright holders and welcomes correspondence from those they have been unable to contact.

ISBN 13: 978-1-315-89684-7 (hbk)
ISBN 13: 978-1-351-07594-7 (ebk)

Visit the Taylor & Francis Web site at http://www.taylorandfrancis.com and the
CRC Press Web site at http://www.crcpress.com

PREFACE

This book is the result of many years experience related to the preparation and use of precautionary labels for chemical containers. The experience includes both an academic and industrial environment. I have tried to provide both regulatory interpretations and practical methods for meeting these mandates.

The original idea of writing this book came from my publisher. The contents evolved from the original design as the project become more mature. It is my sincere hope that the final product helps everyone who writes or uses precautionary labels to do their job better.

Many companies have cooperated and given me permission to reproduce their labels and quote from their materials. Some of these companies are clients of my employer with whom we have had long-term, enjoyable relationships. Others are companies I contacted because their labels illustrate specific information or techniques. I appreciate the cooperation extended by each of these companies and their employees with whom I dealt.

Some of the material is based on case histories taken from the files of CHEMICAL SAFETY ASSOCIATES, Inc. The facts of these cases are accurately reflected in the reports; sufficient information has been changed to protect the confidentiality of those involved.

Lastly, to all of those, including my publisher, who have helped and tolerated delays, typos, and general production problems, I say "Thank you for your patience."

Neal Langerman, 1993

THE AUTHOR

Neal Langerman received his B.A. in Chemistry from Franklin and Marshall College in Lancaster, Pennsylvania and his Ph.D. in Physical Chemistry and Thermodynamics from Northwestern University in Evanston, Illinois. In 1981, after 15 years of teaching chemistry and biochemistry, he helped establish Chemical Safety Associates, Inc.; the company he now works for. Dr. Langerman consults with many companies in general industry as well as in the semiconductor industry to improve chemical handling practices, develop emergency response teams, and upgrade industry safety practices. His areas of expertise include chemical safety, environmental protection, regulatory compliance, and training. Dr. Langerman is a Registered Environmental Assessor and a Certified Environmental Inspector. He is experienced with air modeling and environmental remediation. He provides litigation support and expert testimony related to Failure-to-Warn litigation and prepares precautionary labels for chemical containers and Material Safety Data Sheets for chemical products. Dr. Langerman has been providing safety, health, and environmental consultation to the semiconductor industry since 1980. Beginning in 1989, Dr. Langerman has been involved in a project involving a worldwide review of all arsine exposure accidents.

TABLE OF CONTENTS

LIST of TABLES

LIST of FIGURES

INTRODUCTION

Chapter 1

PRECAUTIONARY LABELS FOR CHEMICAL CONTAINERS

INTRODUCTION

Chemicals have been used since the earliest recorded history in the form of dyes, inks, paints, perfumes, as well as in manufacturing. The concept of an industrial chemical as understood at the end of the 20th century dates to the very beginning of the century. While chemists practiced their art in academic settings throughout the 18th and 19th centuries, it wasn't until the onset of mass manufacturing methods that chemicals began to be used in what is today called the "industrial setting."

Containers of chemicals have been identified in various ways throughout recorded history. In general, the form of identification was for the convenience of the person who developed and used the label. It wasn't until 1945[1] that the format of the industrial chemical label began to evolved.

The chemical label is currently **the primary form of hazard communication.**[2] It is aimed at the person who uses the chemical and is designed to remind that person of the potential hazards associated with the chemical. This fact is all-to-often overlooked in the work place. Precautionary labels should only be used to the extent necessary to warn of appropriate hazards. The warnings should be based on the inherent properties of the chemical or mixture *as well as* potential hazardous exposures resulting from reasonably foreseeable occupational use, handling, storage, and misuse. The language used on a chemical label should contain instructions that, if followed, are sufficient to assure that no injury or illness will occur as a result of handling the chemical.

1 MANUFACTURERS CHEMISTS ASSOCIATION (MCA) Warning Labels (Manual L-1), 1945

2 *Federal Register*, 48 No. 228, 53300 - 53301, November 25, 1983

The chemical label is also a regulatory compliance method. The U.S. Occupational Safety and Health Administration (OSHA), the Canadian Safety Administration (CSA), the U.S. Environmental Protection Agency (EPA), and the U.S. Consumer Product Safety Commission (CPSC) all have issued primary standards for communication of potential hazards via labels.

The label, in addition, is designed to protect the label preparer from possible litigation. The concept of "adequate warning" has become central to the entire body of civil law generally referred to as "Failure-to-Warn" litigation. A well-designed label, which complies with the myriad of regulatory demands, meets human factor considerations for readability, lay-out, and accessibility, significantly deters aggressive civil litigation.

Finally, the label is a form of product recognition. Chemical manufacturers adopt a uniform format for their product line with the objective of maintaining customer loyalty in exactly the same manner as a personal care product manufacturer does.

By the standards of the end of the 20th century, a label serves a myriad of purposes and must be designed to satisfy them all.

PURPOSE

Chemical labels are a primary form of hazard communication. For communication to occur, however, all parties trying to communicate must agree on the form of communication and the meaning of the words and symbols used. It is the intent of this book to provide readers with a common understanding of the label as a form of hazard communication and to develop and present a common vocabulary.

Chemical labels must meet a variety of regulatory constraints depending on the nature of the labeled product. Thus, pesticides must be labeled as defined by the U.S. Environmental Protection Agency (EPA) regulations promulgated under the Federal Insecticide, Fungicide, and Rodenticide Act (FIFRA); industrial products must be labeled as defined by U.S. Occupational Safety and Health Administration (OSHA) regulations or the Canadian Safety Administration (CSA) regulations; products in transportation must be labeled as required by the Department of Transportation (DOT); consumer products must be labeled as required by the Consumer Product Safety Commission (CPSC); and, some specific classes of chemicals must be labeled for minimizing environmental risks, as defined again by the EPA.

RISKS AND HAZARDS

The function of a chemical label is to communicate information about chemical hazards and potential risks. Thus, it is essential to agree upon and understand the meaning of both terms. A *chemical hazard* is a condition or property that has the potential to do physical harm or cause illness (acute or chronic). Acute hazards have the potential to cause harm within a short time. Some common examples of acute hazards include skin corrosivity, flammability, and extreme toxicity. Chronic (or delayed) hazards are those whose potential to cause adverse affects only is expressed over a prolonged period of (often low) exposure. Chemicals that cause cancer, induce fetal abnormalities, or cause irreversible damage to organs such as the liver, kidneys, or brain are illustrations of such chronic effects.

This book includes a rather extensive explanation of terms and pictograms commonly found on chemical labels. The purpose of these explanations is to provide a commonality of language associated with precautionary labels.

ELEMENTS OF A CHEMICAL LABEL

A chemical label is the user's primary source of information regarding the product. Figure 1 presents a warning label from a typical household chemical.

FIX-IT-FAST DRAIN CLEANER

PRECAUTIONS
READ BEFORE USING
DANGER! POISON!

Harmful or fatal if swallowed - Corrosive: causes severe burns. Contains sodium hydroxide (caustic soda-lye). Keep lye and solutions of lye away from eyes, skin, clothing; do *not* get on you. Always keep lye out of reach of children. Do not transfer lye to, or store in another container. Store on high shelf or in locked cabinet.

FIRST AID - **SKIN** - Flush with water for fifteen minutes. **EYES** - Immediately hold face under running water for twenty minutes with eyes open, by force if necessary. Then cover with clean dressing or sheet. **IN MOUTH OR IF SWALLOWED** - *Do Not* induce vomiting. Clear mouth. Give (drink) large quantities of MILK or water. Then seek medical attention immediately. **TRANSPORT PERSON TO NEAREST EMERGENCY CENTER and CALL PHYSICIAN.**

FIGURE 1 Example of a consumer product label.

Examination of this label will determine that it has the elements needed for user safety. The product name is clearly identified. The label also has directions for use. The precautionary statements include the "signal word" **DANGER!** and the added attention-getting term **POISON!**, with the associated "Skull and Crossbones" pictogram.

The label tells the user explicitly what can happen in case of exposure **"harmful or fatal if swallowed. Corrosive: causes severe burns."** The label gives very specific steps the user should take to avoid exposure and what to do in case of exposure.

This information is adequate for home chemical use. Additional detail may be needed in the industrial work place. The history of attempts at developing a standard for labeling chemicals in the workplace dates back to the "Labels and Precautionary Information Committee"[3] of the Chemical Manufacturers' Association (as it is now known) in 1944. This committee published its recommendations in 1946, and these have been the *de facto* standard for labels on most chemical containers. This standard has been formalized and adopted by the American National Standards Institute (ANSI) and is now recognized as the ANSI Labeling Standard Z129.1-1988.

The standard recommends the following information on a chemical product label:

1. **Identity of product and hazardous constituents.**
2. **Signal word -DANGER!, WARNING!, or CAUTION!.**
3. **Statement of hazard.**
4. **Precautionary measures.**
5. **Instructions in case of contact or exposure.**
6. **Antidotes.**
7. **Notes to physicians.**
8. **Instructions in case of fire, spill, or leak.**
9. **Instructions for container handling and storage.**
10. **Other useful information.**

It is the intent of this book to provide the user with the knowledge and skill to understand and use this information correctly. This book will focus on labels for chemical containers, typically of 110 gallons or less. Other topics related to labels will

[3] See Footnote 1

be addressed as is appropriate.

ORGANIZATION OF THIS BOOK

This book has been organized to allow the reader to focus on those specific sections of most immediate importance. Chapters 2 and 3 address the regulations related to chemical labels and to potential litigation resulting from inadequate labels. Chapters 3 through 6 are aimed at the label preparer. These chapters address the many considerations related to the actual words that be placed on the label. Chapter 7 is unique in that it addresses those special purpose labels that been developed or are mandated to convey a particular warning. Finally, chapters 8 and 9 address label and Material Safety Data Sheet (MSDS) use. MSDS use has been included because the label and the MSDS are complimentary aspects of the communication process, and both must be used to provide sufficient information to handle chemicals safely and effectively.

The reader of this book is assumed to have a basic familiarity with industrial and laboratory chemicals. The illustrations selected are based on the need to emphasize particular aspects of the label. As readers use this book, their awareness and understanding of chemical labels will increase dramatically.

DANGER! • WARNING! • CAUTION!

REGULATIONS RELATED TO CHEMICAL LABELS

Chapter 2

PRECAUTIONARY LABELS FOR CHEMICAL CONTAINERS

INTRODUCTION

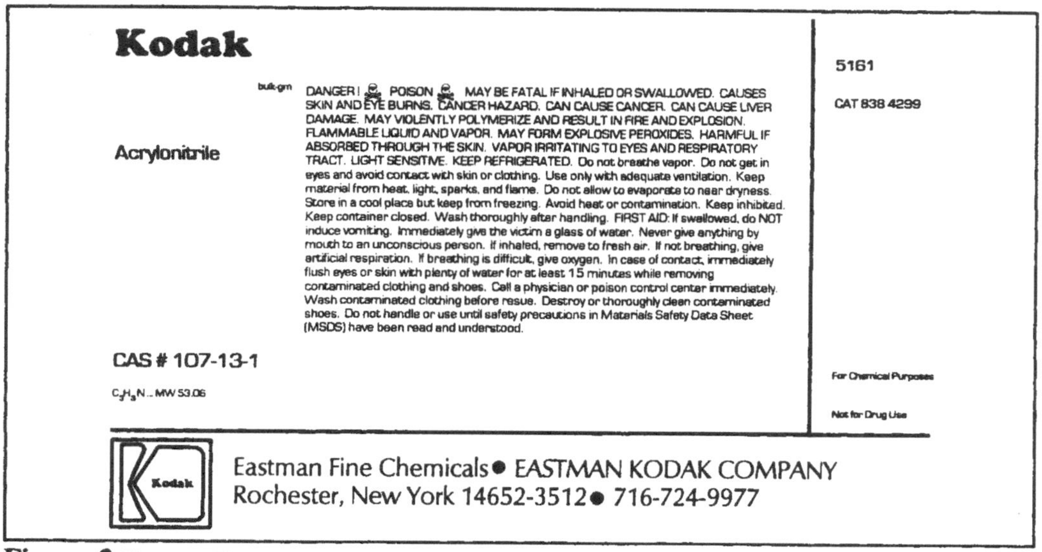

Figure 2 Eastman Kodak Company label for acrylonitrile.

Precautionary labels reflect the impact of governmental regulations on society. The occurrence, design, and content of a label is extensively mandated by various regulatory agencies. In order to understand the communications content of a label, it is necessary to appreciate the impact of regulations on labels.

Precautionary labels are primarily designed to communicate information regarding potential hazards and, thus, to minimize the possibility of injury or illness. Figures 2 through 4 present examples of labels for several industrial products. These illustrate many of the regulatory issues that will be described in this chapter.

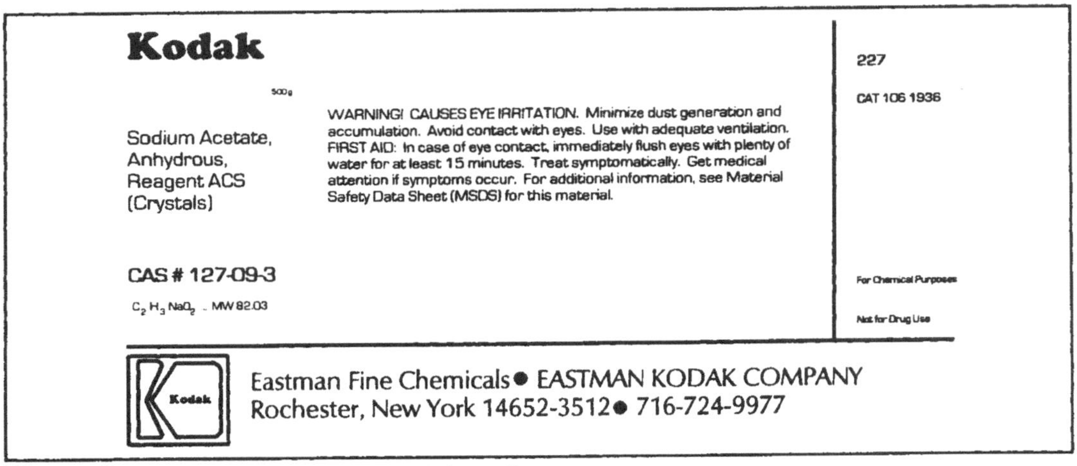

Figure 3 Eastman Kodak Company label for sodium acetate.

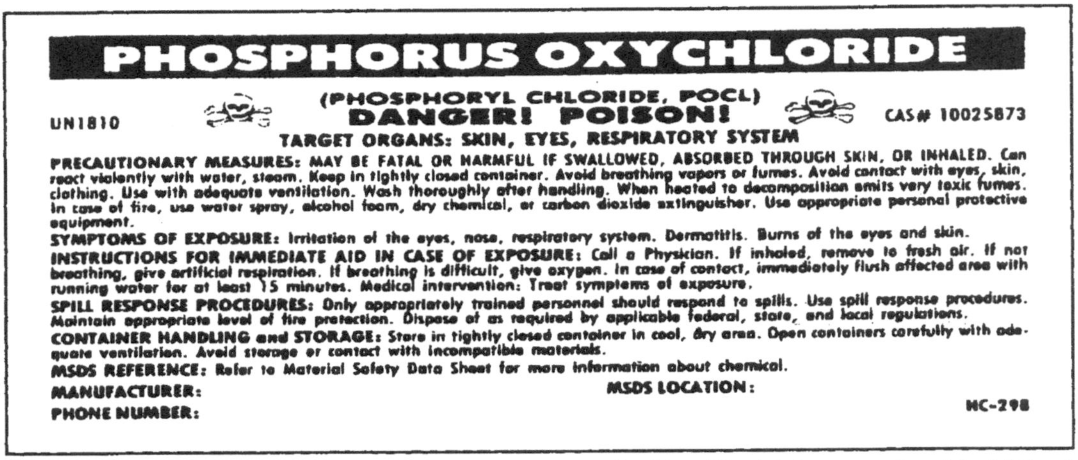

Figure 4 Secondary container label for phosphorous oxychloride. The label was produced by National Marker Company.

Examination of these labels indicates a number of common features: the product is clearly identified, and an appropriate signal word is associated with the product name. The primary potential sources of injury are identified and measures are presented to avoid these primary hazards. First-aid measures and the most significant handling and storage information is given. Finally, the label refers the end-user to the MSDS and gives the chemical manufacturer's name, address, and telephone number. Not surprisingly, each of these items can be traced to either an OSHA or Industry standard for label content.

OSHA REGULATIONS

The OSHA "Hazard Communication Standard" defines the specific requirements for precautionary label content, availability, and usage in the workplace. The standard was originally issued in November, 1984, and its coverage was expanded in August, 1988. The standard requires that the label contain *at a minimum* the identity of the material, statements concerning the major hazards, and the name and address of the manufacturer.

The Hazard Communication Standard, however, also cross-references the American National Standards Institute (ANSI) "Standard on Industrial Labels."[4] This industry consensus standard requires very specific information for an industrial label. These items include:

[4] OSHA is in fact referring to the American National Standard for Hazardous Industrial Chemicals - Precautionary Labeling; ANSI Z129.1-1988.

1. Identity of product and hazardous constituents.
2. Signal word - DANGER!, WARNING!, or CAUTION!.
3. Statement of hazard.
4. Precautionary measures.
5. Instructions in case of contact or exposure.
6. Antidotes.
7. Notes to physicians.
8. Instructions in case of fire, spill, or leak.
9. Instructions for container handling and storage.
10. Other useful information.

Not all items will be needed for all chemicals. Item 7, Notes to physicians, will only be required when such information is available. Labels for cyanide-containing compounds, for example, should have a reference to using amyl nitrite inhalers as an immediate antidote to cyanide poisoning.

OSHA specifically requires in 29 CFR 1910.1200(f) that the manufacturer, importer, or distributor shall ensure that each container leaving the workplace is labeled, marked, or tagged with the following information:

1. Identity of the hazardous chemical(s)
2. Appropriate hazard warnings
3. The name and address of the chemical manufacturer or other responsible party.

This wording is rather general, and the guidance provided by ANSI is essential. An additional item that OSHA required to appear on a label as a result of the Hazard Communication Standard is information regarding the "target organs." OSHA did not require this information in its original standard. Rather, the Secretary of Labor delineated label requirements in the OSHA Instruction to its Field Agents[5] which included the requirement that target organ information appear on the label. This ruling held until 1992. The Occupational Safety and Health Review Commission ruled [6] that the target organ requirement was not reasonable "given the original intent of the Standard not to require target organ warnings on labels". At the time of publication of this book, OSHA had not appealed this ruling.

OSHA also addressed situations where the shipping container is a truck, tank car, rail car, or the like by establishing a requirement that the appropriate label or label information be either posted on or attached to the tank or vehicle or attached to the shipping papers. The chemical manufacturer (the shipper) choosing to attach labels or label information to such a bulk container should communicate the location of this information prior to shipment. **Manufacturing users must ensure that their employees are aware of the label warning before potential exposure to incoming chemicals occurs!**

OSHA established the Material Safety Data Sheet (MSDS) as the method for the chemical manufacturer, importer, or distributor to inform the downstream user of all potential hazards associated with the chemical. Since the imposition of this requirement

5 OSHA Instruction CPL 2-2.38

6 15 OSHC 1497, February 7, 1992. *Secretary of Labor v. American Cyanamid Co.* After this book was prepared, the ruling regarding target organs was reversed. This reversal is published in *Secretary of Labor v. American Cyanamid Co., 5 F.3d 140 (6th Cir. 1993)*.

in 1984, the content of MSDS's has increased. They have become, in many cases, detailed technical reviews of the physical, health, and environmental concerns for a product. The MSDS can be in excess of 10 pages in length. Chapter 9 of this book examines the content of an MSDS in detail.

It is not *necessarily* appropriate to warn of every hazard listed on the MSDS on the label. The MSDS is to address essentially everything that is known about the chemical. The selection of hazards highlighted on the label will involve judgment regarding each hazard reported on the MSDS. This does not mean, however, that only acute hazards are to be reported on the label. Nor does it mean that well-substantiated hazards can be left off of the label because they appear on the MSDS.

The purpose[7] of labels under the Hazard Communication Standard is clear. **Labels provide an immediate warning to employees and link the chemical product to the more detailed information available on the MSDS.** Because of this clear purpose, the content of an in-plant label will be different from a shipping label.

The Hazard Communication Standard requires that every container shipped off-site by the manufacturer, distributor, or importer (which contains hazardous materials as defined by OSHA) has an "appropriate hazard warning" placed on it. The specific information must address both "physical" and "chemical" hazards. The signal words, **DANGER!**, **CAUTION!**, and **WARNING!** are adequate warnings in establishments where training documents employee understanding of the hazards associated with such terms. In other circumstances, such as during shipment, such signal words are **clearly inadequate**. The OSHA definition of "hazard warning" states "*the warning must convey*

[7] See Footnote 2

the hazard of the chemical." If it is well known that inhalation may cause lung damage (hydrochloric acid, for example) then "lung damage" is the hazard, NOT inhalation. In situations for which specific target organ information is not available, then a more general warning is appropriate. For example, if the only information available are LC_{50} data, then the no specific hazard can be listed, and the general warning "harmful if inhaled" may be appropriate.

In-house Labels

OSHA also recognized that a variety of useful labeling systems existed at the time the Standard was issued the "Hazardous Material Information System" (HMIS) is probably the best known and others continue to be developed. Any of these systems can be used, such as that illustrated in Figure 5, as long as the intent of the Standard is met: hazards associated with the chemical are conveyed to the employee. OSHA encourages employers to demand that hazardous effect and target organ information be on labels for all incoming products. OSHA further recommends that employers who choose to use an in-plant labeling system do so *in addition to* the chemical manufacturer's labeling system and allow the original contain label to remain in place.

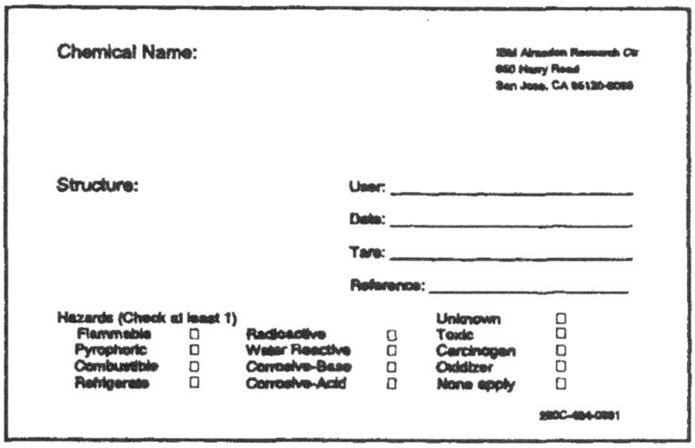

Figure 5 IBM Almaden Research Center in-house label. This label is placed on all chemical containers when received at the facility.

OSHA requires that any portable container which is intended for the immediate use of the employee (ie. the container is NOT used during more than one shift, or does

not leave the control of the person who filled it) must be adequately labeled. OSHA recognized
that a trained employee, familiar with the information content of a labeling system used at the place of employment, would not need the detailed and specific information found on a manufacturer's label. Thus, OSHA allowed the use of a wide variety of in-house labels. Many of these labels were derived from the HMIS labels.

In general, these HMIS-based systems use color-coded fields to represent particular hazards and rate the intensity of the hazard of a scale of 0 (minimum or no hazard) through 4 (extreme hazard). In addition, in-house labels, such as that shown in Figure 6 or Figure 7 (Spanish version) include pictograms illustrating specific hazards, control measures, or protective equipment. Table 1 presents a detailed explanation of the interpretation of the HMIS-based numerical values.

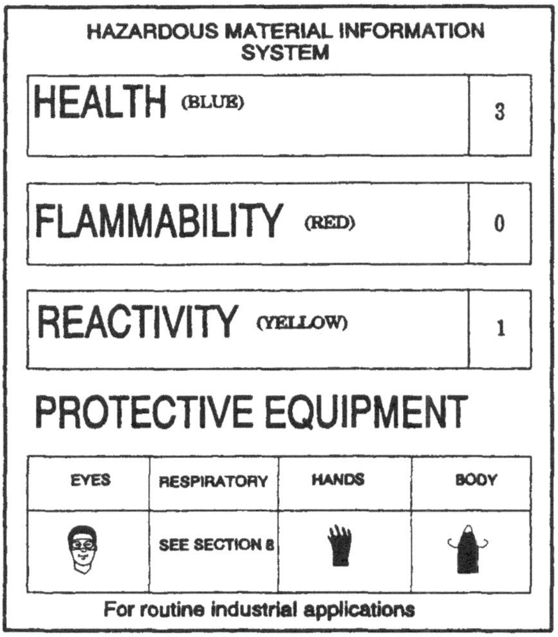

Figure 6 Example of generic in-house label. This label is based on the original HMIS Label system.

TABLE 1
HAZARDOUS MATERIAL INFORMATION SYSTEM
CHEMICAL RATING SYSTEM
CRITERIA

HAZARD RATINGS		HEALTH	FLAMMABILITY	REACTIVITY	TARGET ORGAN EFFECTS
RANK	DESCRIPTION				
4	SEVERE	Deadly. Even the slightest exposure would be life threatening. Specialized clothing for these materials must be worn.	Flash point < 73°F. This substance is very flammable, volatile, or explosive. Extreme caution is needed when handling or storing this material.	May detonate. Substances that are readily capable of detonation or explosion at normal temperatures and pressures.	Severe caustic and corrosive effect on tissues. Toxic effect on blood, liver, kidneys, and/or nervous system. Special protective clothing needed against specific hazards. Skin should not be exposed.
3	SERIOUS	Extreme Danger. Serious injury would result from exposure to this substance. Do not expose any body surface to these materials. Full protective measures should be taken.	Flash point < 100°F. Flammable, volatile, or explosive under almost all normal temperature conditions. Exercise great caution in storage or handling.	Explosive. Substances that are readily capable of detonation or explosion by a strong initiating source, such as heat, shock, or water.	Highly toxic and irritating to mucous membranes, lungs, eyes, skin, and digestive system. Full protective clothing needed. Skin should not be exposed.
2	MODERATE	Dangerous. Exposure to this substance would be hazardous to health. Protective measures are indicated.	Flash point < 200°F. Moderately heated conditions may ignite this material. Caution should be used when handling.	Unstable. Violent chemical reactions are possible at normal or elevated temperatures and pressures. Potentially explosive or violent reaction when mixed with water.	Toxic. Irritant to mucous membranes, skin, lungs, kidneys, liver, and respiratory organs. Provide adequate supplied air respiratory protection.
1	SLIGHT	Slight hazard. Irritation or minor injury would result from exposure to this substance. Protective measures are indicated.	Flash point > 200°F. This material must be preheated to ignite.	Normally stable. Substances that may become unstable at elevated temperatures and pressures or when mixed with water.	Irritating vapor and fumes in contact with skin, eyes, and mucous membranes.
0	MINIMAL	No hazard. Exposure to this substance offers no significant risk to health.	Will not ignite.	Stable. Substances are stable when exposed to heat, pressure, or water.	Slight irritant. Protection for eyes, skin, and upper respiratory system needed.

OSHA does not regulate labeling of very small containers, such as found in laboratories. OSHA does not comment on this issue in the Laboratory Standard.[8] Thus, the employer is forced to develop a procedure for each facility. Laboratories in particular, have a serious problem, since many sample containers of various sizes with a diversity of chemical contents exist. OSHA did not comment on this issue directly in the Laboratory Standard[9] but chose to use the same wording with regard to labels as found in the Hazard Communication Standard. OSHA did specifically state that the mechanism of laboratory labeling was left to the employer to implement as part of the Chemical Hygiene Plan.[10] The labels shown in Figures 5 - 7 are available in sizes to fit almost any container. In the case of the label shown in Figure 5, company policy requires this label on every container, regardless of the presence of an original label *and* regardless of size.

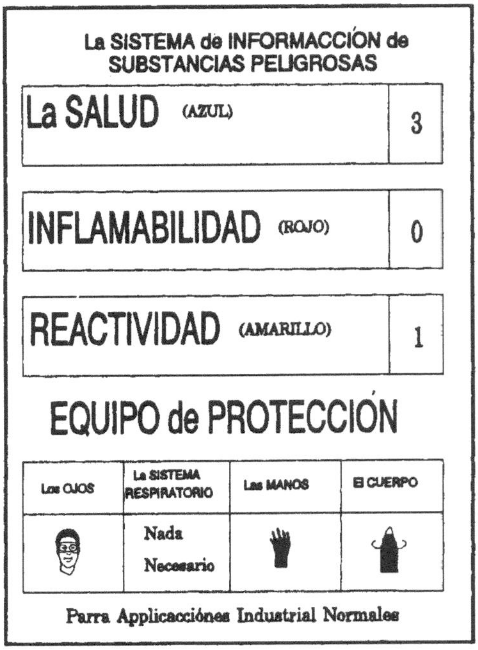

Figure 7 Spanish version of an in-house label for secondary containers. This label, produced by National Marker, combine pictograms and numbers to convey precautionary information.

8 29 CFR 1910.1450

9 29 CFR 1910.1450

10 55 Federal Register 3325, January 31, 1990

Carcinogen Labeling

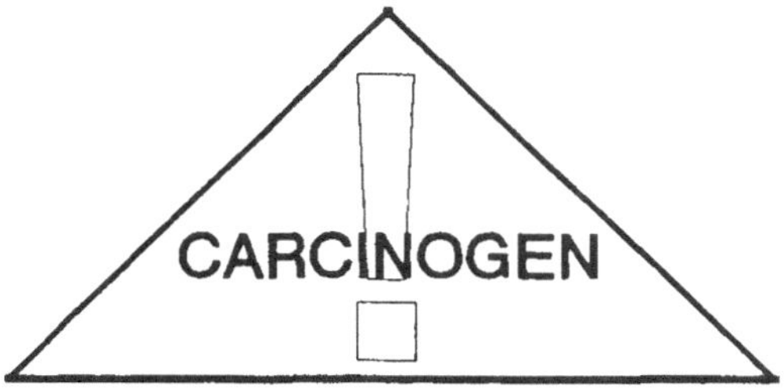

Figure 8 Additional label that may be placed on containers whose contents require a carcinogen warning.

OSHA specifies that all carcinogens present in a mixture in concentrations greater than or equal to 0.1% must be listed on the MSDS. OSHA does not provide an equivalent specification for labels. OSHA allows employers to assess whether a carcinogen warning must be on a label. Refer to the wording used on the Acrylonitrile (Figure 2) label for an example of a carcinogen warning. The label clearly states "... **CANCER HAZARD. CAN CAUSE CANCER.** ..." OSHA also, however, places responsibility to defend any assessment on the employer. Such assessments must be well documented.

OSHA considers a chemical to be a carcinogen if it has been listed as a "positive or suspected carcinogen" in one (or more) of the following references:

National Toxicology Program (NTP) Annual Report on Carcinogens

International Agency for Research on Cancer (IARC) Monographs

29 CFR 1910, Subpart Z, Toxic and Hazardous Substances

OSHA also states that the existence of one valid, positive study indicating carcinogenic potential is sufficient basis for notation on the MSDS. If such studies include positive human evidence of carcinogenicity, then the label must contain a hazard warning on carcinogenicity.

OSHA has instructed its Field Compliance Officers that the label must contain a carcinogenicity warning under the following conditions:

1. *The National Toxicology Program (NTP) has listed the chemical as being "known to be carcinogenic";*
2. *The NTP has listed the chemical as being "reasonably anticipated to be carcinogenic";*
3. *IARC has listed the chemical as a "Category 1 ... There is sufficient evidence to support a causal association between the exposure and cancer";*
4. *IARC has listed the chemical as a "Category 2A ... exposures for which there is at least limited evidence of carcinogenicity to humans";*
5. *IARC has listed the chemical as a "Category 2B ... there is inadequate data in humans, but there is sufficient evidence (of carcinogenicity) in animals";*
6. *One of the "OSHA Known Carcinogens":*

Asbestos	**N-nitrosodimethylamine**
4-nitrobiphenyl	**Vinyl chloride (and polyvinyl chloride)**
α-naphthylamine	**Inorganic arsenic**
Methyl chloromethyl ether	**Lead**
3,3'-dichlorobenzidine (and its salts)	**Benzene**
Bis-chloromethyl ether	**Coke oven emissions**
β-naphthylamine	**Cotton dust**
Benzidine	**1,2-dibromo-3-chloropropane**
4-aminodiphenyl	**Acrylonitrile**
Ethyleneimine	**Ethylene oxide**
β-propiolacetone	**Formaldehyde**
2-acetylaminofluorene	
4-dimethylaminoazobenzene	

The carcinogen warning can be included in the body of the label (See Figure 2) or as a separate label, as shown in Figure 8.

Label Content

In the United States, the combination of the ANSI Z129.1-1988 and OSHA's Hazard Communication Standard defines the content and usage of chemical warning labels. Label content and usage has become an integral part of the industrial work site. Failure to use adequate labels eventually leads to regulatory citations and/or accidents.

The ANSI Standard provides specific guidance for the preparation of a precautionary label. The items listed previously to be included on a label are taken directly from the ANSI Standard.

Identification - Shall be adequate to permit the selection of proper action in case of exposure. For mixtures, the chemical names of the components contributing substantially to the hazards shall be included as part of the label. If the components are part of a trade secret, an appropriate procedure must be available to provide information in case of an emergency and appropriate precautionary labeling must be used.

Signal Word - Shall indicate the relative degree of severity of an immediate hazard in diminishing order: **DANGER!**, **WARNING!**, **CAUTION!**. When a chemical has more than one signal word, only the signal word corresponding to the class of greatest hazard shall be used. When differentiating immediate from delayed hazards by the optional use of a signal word for *delayed* hazards, a word other than **DANGER!**, **WARNING!**, or **CAUTION!** should be used.

For highly toxic materials, additional attention to the potentially severe harm that can result from exposure will be obtained from the use of the word **POISON!** and the skull and crossbones symbol, normally placed directly above the **Instructions in Case of Contact or Exposure**. When used, this legend shall not replace the signal word, which always will be **DANGER!**.

Statement of Hazard - Shall give notice of the hazard(s) that are present in connection with the customary or reasonably foreseeable handling, use, or misuse of the product. For chemicals possessing more than one such hazard, an appropriate statement for each immediate hazard must be included. Warning of immediate hazards generally precedes warnings for delayed hazards on the label. Examples of these statements and of the next several label sections are given in Table 2.

Precautionary Measures - Shall supplement the **statement of hazards** by briefly providing measures to be taken to avoid injury from the hazard(s). In general, reference to specific **Target Organ(s)** should be included. As a result of the Canadian Workplace Hazardous Material Information System (WHMIS) regulations (see below), a statement as to the availability of the MSDS also should be included on the label.

Instructions in Case of Contact or Exposure - Shall be included where the results of contact or exposure warrant immediate treatment (first-aid) and where simple remedial measures may be taken before professional medical assistance is available. Such instructions shall be limited to actions that can be taken by individuals without special training.

First-Aid - When simple remedial actions can be taken that will avoid or lessen severe injury, these should be included in this section. This section need not repeat statements made previously. When specific **antidotes** are known and are of such a type as can be administered by a person without training, they shall be included on the label in the **First-Aid** section under a subheading **Antidotes**. **Antidotes** or **Recommended Medical Practices** that should be administered by a qualified health care provider shall be included on the label with the caption: **Notes to Physicians**.

Instructions in Case of Fire, Spill, or Leak - Shall be included when applicable to provide persons who handle the chemical during shipment and storage with appropriate instructions for confining and extinguishing fires and for cleaning up spills or leaks. The instructions shall be as brief and simple as possible and advise the use of suitable material for control.

Instructions for Container Handling and Storage - Shall be included for those chemicals requiring special or unusual handling and storage procedures.

Precautionary label statements must be selected to provide adequate warning, but must not *overwarn*. The information presented on the label must be readable and, therefore, must fit on the space available. The amount of text that can be placed on a label is generally limited by the size of the container. An example of a label design is shown in Figure 9.

Table 2 presents a summary of the elements that which should appear on a chemical label. The table also presents examples of the language that may be selected to address each label element.

PRODUCT NAME or IDENTIFICATION
Identity of Hazardous Component(s)

SIGNAL WORD
STATEMENT OF IMMEDIATE HAZARD(S)

Target Organs

Statement of Delayed Hazards

Precautionary Measures

Availability of MSDS

Instructions In Case of Contact or Exposure

First Aid Statements and Antidotes

Notes to Physicians

Instructions in Case of Fire, Spill, or Leak

Instructions for Container Handling and Storage

Additional Useful Information

Net Contents

Name and Address of Company

Figure 9 Typical design for precautionary label.

CANADIAN WORKPLACE HAZARDOUS MATERIAL INFORMATION SYSTEM (WHMIS)

The Canadian equivalent to the OSHA Hazard Communication Standard is WHMIS. This standard affects almost all Canadian businesses. WHMIS became law on January 20, 1988, and it became effective in October, 1988. The enabling legislation was Canadian Bill C-70. This standard defines label and MSDS content and usage in Canada. Many similarities and differences exist between these two standards. Perhaps the most significant difference, however, is that Canadian law requires that warning labels and MSDSs be available in the workplace in *both* English and French. An example of a Canadian WHMIS label is shown in Figure 10.

WHMIS is a national standard that was developed as a consensus effort on the part of government, labor, business, and the community. It establishes a uniform set of requirements to identify the hazards of materials produced, sold to, imported into, or used in Canadian workplaces. Such hazardous materials are referred to as *Controlled Products* under the Hazardous Products Act and Regulations. Since controlled products appear in almost all areas of commerce, the Act affects almost all manufacturers, importers, and distributors of such controlled products as well as all end-users of the hazardous materials.

TABLE 2

EXAMPLES OF LABEL STATEMENTS

PRIMARY HAZARD	PRECAUTIONARY STATEMENTS	INSTRUCTIONS IN CASE OF EXPOSURE	FIRST-AID	INSTRUCTIONS IN CASE OF FIRE, SPILL, OR LEAK	INSTRUCTIONS FOR HANDLING & STORAGE
Extremely Flammable	Keep away from heat, sparks, & sources of *ignition*	Rinse affected area with running water for at least 15 minutes. If inhaled, remove victim to fresh air. If not breathing, provide CPR.	Victim(s) of chemical exposure must obtain medical assistance. Take victim and MSDS for product to medical care immediately.	Only trained responders should attempt to clean-up a chemical spill. Control source, control spread of material. Dispose of residue properly.	Store in a cool, dry location. Store only with other chemicals which are compatible.
Water Reactive Flammable Solid	Keep away from water, heat, sparks, and sources of ignition.				Do not store in a location protected by automatic water-based fire suppression equipment.
Fuming Corrosive	Avoid getting material in you or on you. Do not breathe vapors. Do not allow chemical to come in contact with skin, eyes, or clothing.				Store in a cool, dry location. Store only with other chemicals which are compatible.
Highly Toxic	Use in a well ventilated location. Avoid breathing vapors. Do not get chemical on skin, in eyes, or on clothing. Wash thoroughly after handling.				Access to area should be controlled.
Carcinogen	Causes cancer. Use in a well ventilated location. Avoid breathing vapors. Do not get chemical on skin, in eyes, or on clothing. Wash thoroughly after handling.				Restrict access to this material to authorized personnel only

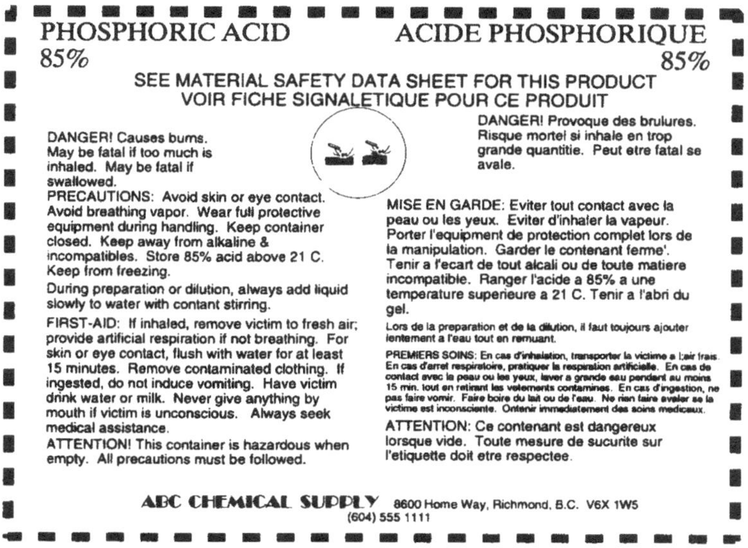

Figure 10 Example of a Canadian WHMIS label for phosphoric acid.

WHMIS is implemented under a combination of federal and provincial legislation. The responsibility for preparation and distribution of labels and MSDS is placed on the supplier; the responsibility for the use and availability of these communication devices is placed on the employer. This division of responsibility is parallel to the U.S. Hazard Communication Standard.

WHMIS provides for trade secret protection in a manner similar to the U.S. standard. The Canadian Government went further, however, than the U.S. Government in that Canada established a Federal "Hazardous Materials Information Review Commission" to determine if trade secret claims are legitimate.

WHMIS specifies what information the "Supplier Label" must contain. Specifically, this includes:

Material Identifier - this is a brand, code, common, generic, chemical, or trade name or a code number and must be identical to the material identifier on the MSDS of the controlled product.

Supplier Identifier - the name and address of the supplier of the controlled product to the Canadian workplace and the name that should also appear on the MSDS. A company logo can also be used in addition to the company name. When a product is resold via a distributor, the distributor's name does not need to appear on the label unless the distributor repackages the controlled product.

Reference to the Material Safety Data Sheet - a statement must appear on the label that the MSDS is available.

Hazard Symbol - the label must contain one or more of the symbols shown in Figure 11, as appropriate for the controlled product.

WHMIS HAZARD SYMBOLS

Classes and Divisions	Hazard Symbol
Class A - Compressed Gases	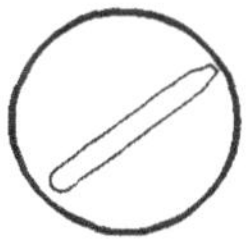
Class B - Flammable and Combustible Material	
Class C - Oxidizing Material	
Class D - Poisonous and Infectious Material	
1. Materials Causing Immediate and Serious Toxic Effects	
2. Materials Causing Other Toxic Effects	
3. Biohazardous or Infectious Material	
Class E - Corrosive Material	
Class F - Dangerously Reactive Material	

FIGURE 11 Examples of WHMIS Hazard Symbols.

Risk Phrases - phrases must be used on the label that indicate the risk(s) associated with the controlled product appropriate to the Class(es) and Division(s) into which it falls.

EXAMPLES OF ACCEPTABLE RISK PHRASES	
SPECIFIC RISK	APPLICABLE LABEL PHRASE
Compressed Gas **Flammable Gas** **Flammable Liquid** **Combustible Liquid**	May form explosive vapor in air
Reactive Flammable Material	Contact with water liberates flammable gas or pyrophoric
Oxidizing Material	Contact with combustible materials may cause fire
Very Toxic Materials	May cause cancer May cause birth defects May cause delayed lung injury May cause irreversible liver damage
Toxic Materials	Eye irritant Respiratory irritant Skin sensitizer
Corrosive Material	Causes burns Causes severe burns Liquid and vapor cause burns

Precautionary Measures - The label must include information on the procedures to be followed when handling, using, or if exposed to the controlled product. The information must be performance based and is the responsibility of the supplier who evaluated the controlled product.

EXAMPLES OF PRECAUTIONARY MEASURES STATEMENTS
Wear eye/face protection. Do not breathe dust. Avoid prolonged or repeated skin contact. Handle and open container with care. Ground containers when in use. Do not empty into drains. Do not drop. Keep container cool and dry. Store in well-ventilated location. Wash or shower after handling.

First-Aid Measures - The label must contain information as to what to do in the event of exposure to the controlled product.

WHMIS also specifies the use of *workplace labels*. When a product that requires WHMIS labeling is transferred to another container in the workplace, the new container must be labeled. If the product is used in the workplace only from the container in which it was shipped, and thus already has a supplier label, the employer is under no obligation to attach a workplace label. Workplace labels must contain the following information:

Product/material identifier

Hazard information

MSDS availability

WHMIS also applies to piping, process tanks, valves, and other parts of process units that contain controlled products. This means that pipes must be labeled as to the contents and direction of flow; valves must be identified as to the material or process controlled; and tanks, mixers, conveyor belts, etc. must be marked with labels or signs, as appropriate.

WHMIS also requires that laboratory samples be "appropriately identified." WHMIS does not clearly specify what constitutes "appropriate" in a laboratory. Good safety practices dictate that all containers be labeled with the name of the contents, the name of the person responsible for the container, and the date it was filled.

WHMIS does not require specific color codes on product labels. WHMIS does require, however, that supplier labels, and laboratory sample supplier labels have a distinctive hatch-mark border. This is illustrated in Figure 10.

WHMIS label requirements are summarized in Table 3.

TABLE 3

SUMMARY OF WHMIS LABEL REQUIREMENTS

Product or Container	Standard or Inner	Small or Inner	Bulk Shipment	Laboratory Sample	Loose Controlled Product	Laboratory Supply House Material	Transferred Product or Workplace Mixes	Produced Products	Exports	Plant Equipment, Pipes, and Valves
Size	> 100 mL	≤ 100 mL	≥ 454 L	≤ 10 kg	N/A	≤ 10 kg	Various	Various	Various	Various
Type Label	Supplier	Reduced Supplier Label	Supplier or Workplace	Laboratory Sample Supplier Label	Workplace Sign or Label	Laboratory Supply House Supplier Label	Workplace Sign or Label	Workplace Label	WHMIS Label Not Required	Identification via Labels, Tags, Signs, Etc.
Product Identifier	Yes	Yes	Yes	Yes	Yes	Yes	Yes	Yes	Workplace Sign or Label if Stored Awaiting Shipment	Yes
Supplier Identifier	Yes	Yes	Yes	Yes	No	No	No	No	No	No
MSDS Availability	Yes	Yes	Yes	No	Yes	Yes, If MSDS Available	Yes	Yes	Yes	Yes and/or Training
Hazard Symbols	Yes	Yes	Yes	No	No	No	No	No	No	No
Risk Phrases	Yes	No	Yes	No	No	Yes	No	No	No	Safe Handling
Precautionary Measures	Yes	No	Yes	No	No	Yes	No	No	No	Safe Handling
First-Aid Information	Yes	No	Yes	No	No	Yes	No	No	No	Safe Handling
Safe Handling Information	Optional	No	Optional	No	Yes	No	Yes	Yes	Yes	Yes
Color Code	Optional	No	Optional	No	Optional	No	Optional	Optional	Optional	Optional
Language	French/ English	French/ English	French/ English	French/ English	Optional	French/ English	Optional	Optional	Optional	Optional
Border	Yes	Yes	No	Yes	No	No	No	No	No	No
Comments/ Notes				Required Laboratory Sample Wording						Additional training Requirements

CONSUMER PRODUCTS

Neither OSHA nor WHMIS regulations apply to consumer products used as normally used in the home. Thus, drain cleaner, which frequently contains sodium hydroxide, is not regulated by either of these standards when sold at retail outlets for consumer, as opposed to industrial, use. In the U.S., warnings on such products are regulated, however, by the U.S. Consumer Product Safety Commission (CPSC). Figure 1 illustrates a typical example of a drain cleaner type product label.

CPSC regulations are published in Title 16 of the Code of Federal Regulations. The information applicable to warning labels is contained in 16 CFR §1500 through 16 CFR §1500.272.

The CPSC regulations apply quite broadly to "hazardous substances," which are defined as:

> *Any substance or mixture of substances which is toxic, corrosive, an irritant, a strong sensitizer, flammable or combustible, or generates pressure through decomposition, heat, or other means, if such substance or mixture of substances may cause substantial personal injury or substantial illness during or as a proximate result of any customary or reasonably foreseeable handling or use, including reasonably foreseeable ingestion by children.*

A hazardous substance, as defined by the CPSC does *not* apply to pesticides regulated by FIFRA, foods, drugs, or cosmetics regulated by the Federal Food, Drug, and Cosmetic Act, or to substances intended for use as fuels when stored in containers and used in the heating, cooking, or refrigeration systems of a house. 16 CFR §1500.3

of this regulation contains a series of definitions that are among the most precise in any federal regulation. Terms such as "extremely toxic" and "significant potential for causing hypersensitivity" are defined in clear operational language, which makes the application of these regulations straightforward. The regulations declare in 16 CFR §1500.12, for example, that charcoal briquettes intended for heating or cooking are a hazardous substances and thus subject to CPSC labeling requirements. In particular, the following wording must appear on the briquette's package:

> WARNING: Do Not Use for Indoor Heating or Cooking Unless Ventilation is Provided for Exhausting Fumes to Outside. Toxic Fumes May Accumulate and Cause Death.

The CPSC regulations refer to the Federal Caustic Poison Act to require the use of the word "**POISON!**" on the label for the following products:

Hydrochloric acid	>10%
Nitric acid	>5%
Oxalic acid	>10% or any salt
Acetic acid	>20%
Potassium hydroxide	>10%
Silver nitrate	>5%
Sulfuric acid	>10%
Carbolic acid (phenol)	>5%
Hypochlorous acid	>10%
Sodium hydroxide	>10%
Ammonia water	>5%

The CPSC regulations require specialized labeling, such as that illustrated previously for charcoal briquettes for products that contain specific chemicals, such as:

Diethylene glycol & mixtures containing	10% (w/w) or more
Ethylene glycol & mixtures containing	10% (w/w) or more
Benzene	>5%
Toluene	>10%
Xylene	>10%
Petroleum distillates	>10%
Methyl alcohol	>4%
Turpentine	>10%

Finally, the CPSC bans products that meet the following descriptions from retail sales in the United States:

Mixtures intended for application to interior floors or walls which are marked "extremely flammable"

Carbon tetrachloride and mixtures containing carbon tetrachloride

Liquid drain cleaners containing 10% (w/w) or more sodium hydroxide and/or potassium hydroxide except if properly packaged as per the requirements of the Poison Prevention Packaging Act of 1970.[11]

Other banned substances include those containing cyanide that may result in a concentration greater than 25 ppm, lead based paint, general use garments containing asbestos, certain fireworks, and certain aerosol products that contain vinyl chloride monomer.

16 CFR §1500.121 defines the "label requirements; prominence, placement, and conspicuousness" for consumer products that contain hazardous substances. The Federal Hazardous Substances Act[12] requires that hazardous substances have certain cautionary statements on their labels. These include:

11 Public Law 91-601; 84 Stat. 1670-74 (15 U.S.C. 1471-76)

12 15 U.S.C. 1261

Signal words

Affirmative statement(s) of the principal hazard(s) associated with a hazardous substance

The common, or usual name, or chemical name of the hazardous substance

The name and place of business of the manufacturer, packer, distributor, or seller

Statements of precautionary measures to follow

Instructions, when appropriate, for special handling and storage

The statement "*Keep out of reach of children*" or its practical equivalent, and

First-aid instructions, when appropriate

The act requires that these cautionary statements appear prominently on the label and be in conspicuous and legible type, in contrast by typography, layout, or color from other printed material on the label.

Central to all of the CPSC regulations resulting from the act is the concept of the Principal Display Panel. This is defined as the "*portion(s) of the surface of the immediate container or any outer container or wrapping which bear(s) the labeling designed to be the most predominantly displayed, shown, presented, or examined under conditions of retail sale.*" The CPSC requires that all cautionary labeling required by the act appear on the Principal Display Panel and on any other container or wrapper. The cautionary information must be generally placed parallel to the base upon which the package rests but exceptions are made for packages for which this will not readily work.

For packages for which the Principal Display Panel is too small for all of the cautionary information, the CPSC allows the label preparer to place the signal word, the statement(s) of principal hazard(s), and the words *"Read carefully other cautionary information on the (indicate location) panel."* If the package has more than one Principal Display Panel, this information must be placed on each such panel. If the Principal Display Panel is a cap or other device that can be readily separated from the main package, then this same information must appear on a prominent panel on the main packaging. Finally, the CPSC requires that when any outer wrapper or packaging obscures the cautionary information on the primary package, that information must be repeated on the outer packaging.

Use of the Skull and Crossbones Pictogram

For highly toxic substances, the label must have the skull and crossbones pictogram, shown in Figure 12, and the word "**POISON!**" in association with other cautionary information. Whenever the word "**POISON!**" is placed on the label, the SIGNAL WORD "**DANGER!**" must be used, unless the CPSC has determined that the word "**POISON!**" replaces the SIGNAL WORD for a particular product.

Figure 12 Example of Skull and crossbones.

Conspicuousness of the Cautionary Information

The CPSC is very specific as to the type size and style. Table 4 presents the sizes of each upper-case letter as it appears on the label. The size of the cautionary labeling shall be reasonably related to the type size of any other printing appearing on the same panel but must meet the minimum sizes presented in the table. The sizes are based on the size of the Principal Display Panel, not necessarily the panel that contains the cautionary information.

TABLE 4

Type Size Requirements for Consumer Labels

Area of Principal Display Panel in square inches	0-2	>2-5	>5-10	>10-15	>15-30	>30
	Type Size in Inches					
Signal Word	3/64	1/16	3/32	7/64	1/8	5/32
Statement of Hazard	3/64	3/64	1/16	3/32	3/32	7/64
Other Cautionary Material	1/32	3/64	1/16	1/16	5/64	3/32

Clearly the requirements for label preparation for a hazardous consumer product are very specific. It is important that the label preparer use the CPSC regulations as a guide for the design and layout of such a label. Because of the complexity of preparing a consumer product label, the Consumer Product Safety Commission will provide "informal" comment on proposed labels.[13]

One area of consumer products has some special labeling requirement - **Art Materials**. Since 1983, the American Society for Testing and Materials (ASTM) has issued an industry consensus standard for labeling art materials with respect to the potential of the material to cause chronic illness. This standard[14] was last revised in 1992. The requirements of this ASTM standard are generally similar to those of ANSI and OSHA, except additional emphasis is placed on chemical constituents that may cause chronic illness or injury. The ASTM label requirements for art materials are described in the following box. The ASTM standard also mandates specific language that should be used to describe various hazards.

---

13 16 CFR 1500.128

14 "Standard Practice for Labeling Art Materials for Chronic Health Hazards," D 4236 - 92, ASTM 1916 Race Street, Philadelphia, PA 19103.

ASTM Label Requirements for Art Materials
Signal Word List of Potentially Chronic Hazards Name of Chronically Hazardous Substances Safe Handling Instructions List of Sensitizing Components Sources of Additional information

TRANSPORTATION LABELS

The U.S. Department of Transportation (DOT) regulates the shipment of hazardous materials. Within the context of this book, the labels and placards that are placed on nonbulk or bulk packaging are "precautionary labels". While it is beyond the scope of this book to present the Hazardous Materials Shipping Regulations in detail, a brief discussion of the hazardous materials labels is appropriate.

Examples of various DOT labels are presented later in this book in Table 7. The labels described in Table 7 and the definitions of the various Hazard Classes are consistent with HM-181, the massive revision of the Hazardous Materials Shipping Regulations adopted by the DOT in 1990. The new regulations were implemented by the DOT in phases, culminating with complete implementation of all phases in 1996. The purpose of the revisions was to harmonize the United States regulations with the United Nations Dangerous Goods Shipping Regulations, generally used throughout the world.

HM-181 divides all Hazardous Materials into nine hazard classes. It further subdivides the hazard classes into divisions. These classes and divisions are explained in the following paragraphs.

Explosive - Class 1. For the purpose of the DOT an "explosive" means any substance or article, including a device, that is designed to function by explosion (i.e., an extremely rapid release of gas and heat) or that, by chemical reaction within itself, is able to function in a similar manner even if not designed to function by explosion, unless the substance or article is otherwise classed by the DOT.

Explosives in Class 1 are divided into six divisions as follows:

Division 1.1 consists of explosives that have a mass explosion hazard. A mass explosion is one that affects almost the entire load instantaneously.

Division 1.2 consists of explosives that have a projection hazard but not a mass explosion hazard.

Division 1.3 consists of explosives that have a fire hazard and either a minor blast hazard or a minor projection hazard or both but not a mass explosion hazard.

Division 1.4 consists of explosive devices that present a minor explosion hazard. No device in this division may contain more than 25 g (0.9 ounce) of a detonating material.

Division 1.5[15] consists of very insensitive explosives. This division is comprised of substances that have a mass explosion hazard but are so insensitive that there is very little probability of initiation or of transition

15 The probability of transition from burning to detonation is greater when large quantities are transported in a vessel

from burning to detonation under normal conditions of transport.

Division 1.6[16] consists of extremely insensitive articles that do not have a mass explosive hazard. This division is comprised of articles that contain only extremely insensitive detonating substances and that demonstrate a negligible probability of accidental initiation or propagation.

Flammable gas Class 2, Division 2.1 (1) For the purpose of the DOT, a "flammable gas" (Division 2.1) means any material that is a gas at 20°C (68°F) or less and 101.3 kPa (14.7 psi) of pressure (a material which has a boiling point of 20°C (68°F) or less at 101.3 Kpa (14.7 psi) that:

(1) Is ignitable at 101.3 Kpa (14.7 psi) when in a mixture of 13 percent or less by volume with air; or

(2) Has a flammable range at 101.3 Kpa (14.7 psi) with air of at least 12 percent regardless of the lower limit.

Nonflammable, nonpoisonous compressed gas Class 2, Division 2.2 This division includes compressed gas, liquefied gas, pressurized cryogenic gas, and compressed gas in solution. For the purpose of the DOT a "nonflammable, nonpoisonous compressed gas" (Division 2.2) means any material (or mixture) that:

(1) Exerts in the packaging an absolute pressure of 280 kPa (41

16 The risk from articles of Division 1.6 is limited to the explosion of a single article

psia) at 20°C (68°F); and

(2) Does not meet the definition of Division 2.1 or 2.3.

Poisonous Gas Class 2, Division 2.3. For the purpose of the DOT, "poisonous gas" (Division 2.3) means a material that is a gas at 20°C (68°F) or less and a pressure of 101.3 kPa (one atm) (a material that has a boiling point of 20°C (68°F) or less at 101.3 kPa (14.7 psi) and that:

(1) Is known to be so toxic to humans as to pose a hazard to health during transportation, or

(2) In the absence of adequate data on human toxicity, is presumed to be toxic to humans because when tested on laboratory animals it has an LC_{50} value not more than 5000 ppm.

Nonliquefied compressed gas. A "nonliquefied compressed gas" means a gas, other than in solution, that in a packaging under the charged pressure is entirely gaseous at a temperature of 20°C (68°F).

Liquefied compressed gas. A "liquefied compressed gas" means a gas that in a packaging under the charged pressure, is partially liquid at a temperature of 20°C (68°F).

Compressed gas in solution. A "compressed gas in solution"[17] is a nonliquefied compressed gas that is dissolved in a solvent.

Cryogenic liquid. A "cryogenic liquid" means a refrigerated liquefied gas having

[17] The only common compressed gas in solution is acetylene, which is dissolved in acetone

a boiling point colder than -90°C (-130°F) at 101.3 kPa (14.7 psi) absolute.

Flammable liquid - Class 3. For the purpose of the DOT, a "flammable liquid" (Class 3) means any liquid having a flash point of not more than 60.5°C (140°F).

Combustible liquid - Class 3. For the purpose of the DOT a "combustible liquid" means any liquid that does not meet the definition of any other hazard class and has a flash point above 60.5°C (140°F) and below 93°C (200°F).

Flammable Solid - Class 4, Division 4.1. For the purposes of the DOT, a "flammable solid" (Division 4.1) means any of the following three types of materials:

(1) Wetted explosives that

(a) When dry are Explosives of Class 1 other than those of compatibility group A, which are wetted with sufficient water, alcohol, or plasticizer to suppress explosive properties; and

(b) Are specifically authorized by name either in the Sec. 172.101 Table or have been assigned a shipping name and hazard class by the Associate Administrator for Hazardous Materials Safety.

(2) Self-reactive materials are materials that are liable to undergo, at normal or elevated temperatures, a strongly

exothermal decomposition caused by excessively high transport temperatures or by contamination; and

(3) Readily combustible solids are materials that

(a) Are solids which may cause a fire through friction, such as matches;

(b) Show a burning rate faster than 2.2 mm (0.087 inches) per second when tested in accordance with DOT procedures;

(c) Any metal powders that can be ignited and react over the whole length of a sample in 10 minutes or less, when tested in accordance with DOT procedures.

Spontaneously Combustible Material - Class 4, Division 4.2. For the purposes of the DOT a "spontaneously combustible material" (Division 4.2) means

(1) *A pyrophoric material.* A pyrophoric material is a liquid or solid that, even in small quantities and without an external ignition source, can ignite within five (5) minutes after coming in contact with air when tested according to paragraph 3.1.1 or 3.1.2, as appropriate, of appendix E to this part.

(2) *A self-heating material.* A self-heating material is a material that, when in contact with air and without an energy supply, is liable to self-heat. A material of this type that exhibits

spontaneous ignition or can ignite if the temperature of the sample exceeds 200°C (392°F) during the 24-hour test period when tested in accordance with DOT procedures.

Dangerous When Wet Material - Class 4, Division 4.3. For the purposes of this chapter, "dangerous when wet material" (Division 4.3) means a material that, by contact with water, is liable to become spontaneously flammable or to give off flammable or toxic gas at a rate greater than 1 liter per kilogram of the material, per hour, when tested in accordance with DOT procedures.

Oxidizer - Class 5, Division 5.1. For the purpose of the DOT, an "oxidizer" (Division 5.1) means a material that may, generally by yielding oxygen, cause or enhance the combustion of other materials. A solid material is classed as a Division 5.1 material if, when tested in accordance with DOT procedures, the mean burning time of the test mixture, is equal to or less than that of the average of the three tests with ammonium persulfate mixture. A liquid is classed as a Division 5.1 material by analogy to existing entries in the Sec. 172.101 Table.

Organic Peroxide - Class 5, Division 5.2. For the purposes of the DOT, an "organic peroxide" (Division 5.2) means any organic compound containing oxygen (O) in the bivalent -O-O- structure and that may be considered a derivative of hydrogen peroxide, where one or more of the hydrogen atoms has been replaced by organic radicals.

Poisonous Material - Class 6, Division 6.1. For the purpose of the DOT, a "poisonous material" (Division 6.1) means a material, other than a gas, that is known to be so toxic to humans as to afford a hazard to health during transportation, or that, in the absence of adequate data on human toxicity:

(1) Is presumed to be toxic to humans because it falls within any one of the following categories when tested on laboratory animals:

(a) Oral Toxicity. A liquid with an LD_{50} for acute oral toxicity of not more than 500 mg/kg or a solid with an LD_{50} for acute oral toxicity of not more than 200 mg/kg.

(b) Dermal Toxicity. A material with an LD_{50} for acute dermal toxicity of not more than 1000 mg/kg.

(c) Inhalation Toxicity. (A) A dust or mist with an LC_{50} for acute toxicity on inhalation of not more than 10 mg/L; or (B) A material with a saturated vapor concentration in air at 20°C (68°F) of more than one-fifth of the LC_{50} for acute toxicity on inhalation of vapors and with an LC_{50} for acute toxicity on inhalation of vapors of not more than 5000 ml/ms; or

(2) Is an irritating material, with properties similar to tear gas, that causes extreme irritation, especially in confined spaces.

Infectious Substance - Class 6, Division 6.2. (a) For the purpose of the DOT, (1) an “infectious substance” (Division 6.2) means a viable microorganism, or its toxin, which causes or may cause disease in humans or animals, and includes those agents listed in 42 CFR 72.3 of the regulations of the Department of Health and Human Services or any other agent that has the potential to cause severe, disabling or fatal disease. The terms "infectious substances" and "etiologic agent" are synonymous. (2) A "diagnostic specimen" means any human or animal material including, but not limited to, excreta, secreta, blood and its components, tissue and tissue fluids, being shipped for purposes of diagnosis. (3) A "biological product" means a material prepared and manufactured in accordance with the provisions of 9 CFR part 102 (Licenses for biological products), 9 CFR part 103 (Biological products for experimental treatment of animals), 9 CFR part 104 (Permits for biological products), 21 CFR part 312 (Investigational new drug application), or 21 CFR parts 600 to 680 (Biologics), and that, in accordance with these provisions, may be shipped in interstate commerce.

Radioactive Material - Class 7 means any material having a specific activity greater than 0.002 microcurie per gram (μCi/g).

Corrosive Material - Class 8. For the purpose of this subchapter, "corrosive material" (Class 8) means a liquid or solid that causes visible destruction or irreversible alterations in human skin tissue at the site of contact, or a liquid that has a severe corrosion rate on steel or aluminum, in accordance with the following criteria:

(1) A material is considered to be destructive or to cause irreversible alteration in human skin tissue if, when tested on the intact skin of an albino rabbit by the technique described in Appendix A to this part, the structure of the tissue at the site of contact is destroyed or changed irreversibly after an exposure

period of 4 hours or less.

(2) A liquid is considered to have a severe corrosion rate if its corrosion rate exceeds 6.25 mm (0.246 inches) a year on steel (SAE 1020) or aluminum (nonclad 7075-T6) at a test temperature of 55°C (131°F).

Miscellaneous Hazardous Material - Class 9. For the purpose of this subchapter, "miscellaneous hazardous material" (Class 9) means a material that presents a hazard during transport, but that is not included in any other hazard class. This class includes:

(1) Any material that has an anesthetic, noxious or other similar property that could cause extreme annoyance or discomfort to a flight crew member so as to prevent the correct performance of assigned duties; and

(2) Any material that is not included in any other hazard class, but is subject to the requirements of the DOT because it meets appropriate definitions for a hazardous substance or a hazardous waste.

These definitions are consistent with those adopted by other transportation-focused regulatory agencies, such as the Canadian Transportation Commission, the International Air Transport Association (IATA), the International Maritime Organization (IMO), and the International Civil Aviation Organization (ICAO).

The labels associated with these Hazard Classes convey specific, though limited,precautionary information, as clearly indicated by the definitions. They are meant to provide warning to persons who might contact the materials in the course of shipment and thus enhance safety. These labels do not convey the breadth or depth of information contained on the other labels discussed in this book.

CONCLUSIONS

A wide variety of regulations impact the design and content of a precautionary label for chemical packages. The manufacturer, importer, or distributor must be careful to comply with this patchwork of regulations to prevent regulatory citations and minimize the potential for litigation.

DANGER! • WARNING! • CAUTION!

FAILURE-TO-WARN

Chapter 3

PRECAUTIONARY LABELS FOR CHEMICAL CONTAINERS

INTRODUCTION

"Failure-to-Warn" is a common allegation in product liability lawsuits, toxic tort claims, and drug cases. Failure-to-Warn was a primary claim in early asbestos cases, and remains today a primary area of law in cases involving negligence related to failure to instruct or train, failure to provide adequate and necessary safety information, or failure to provide postsale risk advisories or retrofits.

A typical Failure-to-Warn case is illustrated with the following example:

> *An experienced welder took an "empty" 55-gallon steel drum home to make into a trash container. He attempted to remove the top of the drum with a cutting torch. The resulting explosion severely burned him, rendering him permanently disabled.*
>
> *The litigation alleged that the company that placed the flammable chemical in the drum failed to provide adequate warning that flammable vapors may be present, even after the drum was "empty." The litigation maintained that the pictogram, shown in Figure 13, should have been permanently affixed to the drum.*

Figure 13 Warning not to use a cutting torch on a container.

> *The jury awarded the plaintiff more than $6 million dollars.*

The chemical manufacturer must look very broadly to define the scope of his or her duty to warn. Complex case law and jury decisions increase the scope and the nature of necessary warnings. The chemical manufacturer must continually evaluate his or her labels to ensure that they provide adequate warning. There are four broad concepts related to Failure-to-Warn that the label preparer must consider when designing a label.

Legal Concepts Related to Failure-to-Warn

1. Those forms of warning that society has adopted and adheres to as customs and practices for an extended period of time are referred to a "Common Law." Most of these common law concepts of label design and usage are addressed in the American National Standard on Precautionary Labeling.

2. Society perceives that chemical manufacturers are in a better position than the chemical user or the public in general to recognize potential chemical hazards associated with their product. Thus, society places a specific duty to warn on the chemical manufacturer. The warnings expected by society must cover both routine use and reasonably anticipated misuse of the product.

3. The prudent manufacturer will recognize and address on the precautionary label clear and current information related to possible environmental damage associated with the use or misuse of the product.

4. The cost of product liability protection, including insurance, defense, and settlements, is continually rising. The prudent manufacturer will implement an on-going program of review of precautionary warning devices.

   In evaluating a precautionary label, the following factors must be considered:

   A. The seriousness of the potential harm posed by the product;

B. The foreseeability of the potential harm;

C. The nature of the warning that should have been given, including factors such as the feasibility of giving the warning and the level of knowledge of the expected product user; and,

D. The extent of the warning - was it sent far enough into the chain of commerce.

Minimization of Failure-to-Warn Litigation Exposure

George A. Peters has written extensively on the topic of Failure-to-Warn. He has prepared[18] a series of short questions designed to assist in presenting a defense to Failure-to-Warn litigation. These questions are valuable in guiding the thinking of the label preparer and reviewer.

1. **WHAT IS THE BASIC OBJECTIVE OF THE WARNING?**

The purpose of the label is appropriately to identify specific hazards, the magnitude of the associated risks, and to describe the methods a person should use to avoid the hazard. The essence of precautionary labels is to provide informed consent or choice behavior. Thus, the label preparer must clearly state the potential hazards and the methods to avoid those hazards, so the user is fully informed of the possible consequences of his actions.

2. **DOES THE WARNING DEAL WITH RESIDUAL RISK?**

---

[18] Peters, George A., Trial Diplomacy Journal, Vol 15, 35-38 (1992)

Precautionary labels should be designed only to warn of hazards that cannot be eliminated by engineering methods. Those hazards that remain after engineering has controlled all possible controllable hazards are referred to as "residual hazards"; the associated risks are called "residual risks."

Polymeric coatings are used on concrete bridges to minimize weather related damage. The coating consists of three components which are mixed just prior to application. The components are the resin, the initiator, and the promoter. When mixed properly, the product performs well. If the initiator and promoter are mixed together, a violent explosion occurs. This is a well-known hazard. Package design must be such that the possibility of this misuse is eliminated. The precautionary label must address any residual possibility of causing this disastrous mixture. This is "residual risk."

Label design goes together with package design. Both must be considered in order to reduce to the greatest extent possible all risk associated with handling the product. There must be a warning on the label of any risk not eliminated.

3. WHAT ARE THE PSYCHOLOGICAL COSTS TO THOSE EXPOSED?

Warnings are only effective if the person at whom the warning is aimed perceives a direct benefit in heeding the warning. The California Proposition 65 warning, when applied to gasoline (which contains the carcinogenic chemical, benzene), is a good example of a situation where the cost-benefit ratio is such that the warning is ignored. The warning gives the vague statement:

> WARNING!
>
> This product contains a chemical known to the State of California to cause cancer.

This warning is perceived as unimportant against the obvious benefit of a full tank of gasoline. Thus, the Proposition 65 warning is ineffective.

When the frequency and severity of the potential injury is perceived to be low, rare, uncommon, or remote, the warning will be ignored. The psychological cost of the adhering to the warning is not adequately balanced by the benefits adherence provides. Thus, for a warning to be effective, the warning must clearly state a net personal benefit for the person to be warned. The greater the personal benefit, the greater the compliance.

4. **IS THE WARNING APPROPRIATELY SUPPLEMENTED?**

OSHA has made it very clear that the purpose of the precautionary label is to provide an immediate warning of the most serious acute and chronic hazards. The Material Safety Data Sheet serves the purpose of providing extensive explanation of these hazards, as well as other, lower risk, hazards. The label serves as the immediate warning device; the MSDS serves as the supplementary device. Thus, the chemical user must have access to the MSDS in order for the label warning to be most effective.

Posted reminders are also effective, if adequately perceived by the chemical user. A sign reminding employees to wear splash goggles posted near a caustic process tank serves to reinforce the warnings on the product label. The Proposition 65 warning placed on the side of a gasoline pump is not effective in alerting the gasoline purchaser to the possible health affects of the product. In order for this to occur, an extensive public education program would be required.

5. HAS THE WARNING BEEN FORMULATED IN A RATIONAL, SYSTEMATIC MANNER?

Proper design of a precautionary label must be performed in a systematic manner. Too often, labels are put together by a product manager, engineer, or attorney. Such labels are not usually the result of a specific design analysis, such as a fault-tree analysis, to identify and quantify potential risks. Nor do such designs usually choose among possible alternatives. Such labels seldom consider the current state of knowledge[19] on label design and implementation. The result of such a haphazard approach to label design is an untested label, which may not serve its intended purpose.

Prevention is much easier than litigation. Proper label design is not difficult. One must be familiar with the current technology and must proceed in a rational, organized manner. The chapter on label design in this book will provide specific procedures to follow.

19 Peters, George, A. "Warnings and Instructions for Packaging Specialists," Hazard Prevention, January/February, 1986

Ryan, Joseph P., "Design of Warning labels and Instructions," Van Nostrand Reinhold, 1991, New York

LABEL DESIGN

Chapter 4

PRECAUTIONARY LABELS FOR CHEMICAL CONTAINERS

INTRODUCTION

The design of a precautionary label must address and balance a variety of considerations: the amount of information versus the maximum size of the container; the information content versus the various applicable regulations; and the literacy and language skills of the chemical user. In this chapter, methods will be presented for preparing a precautionary label. Many specific procedures will be presented, but the experience and judgment of the label preparer is essential.[20]

Getting Started

An industrial precautionary label is required by OSHA and industry standard to have a minimum specific content. The first step in preparing a label is to accumulate the necessary information. The MSDS is one good source of such information. The chemical manufacturer should provide information regarding the intended use and reasonably foreseeable misuses of the product and the packaging. It is important that the label preparer have a structured method for evaluating a label. The use of a check list for this purpose is strongly recommended.[21]

Selecting Appropriate Information

The purpose of the precautionary label is to provide an immediate warning of acute or serious chronic high-risk hazards. Thus, from all of the accumulated information, only those facts that qualify must be selected. This is critical, since selecting too little information could lead to a "Failure-to-Warn" situation, and the alternative could lead to a confusing and cluttered label.

[20] No amount of written material can substitute for actual experience in preparing precautionary labels. Reading this chapter is not sufficient to qualify a person as a label preparer; experience and review of many labels is necessary

[21] Refer to the chapter on Label Evaluation for sample check lists which can be used for this purpose

Product Identification

The selection of the product identification should be simple. Acetone is acetone; nitric acid is nitric acid. Most industrial chemicals, however, are identified by product names or trade names, such as "BL-700." Thus, the first task is to select the exact product identification that will appear on the label. This product identification also must be used as the prime identification on the MSDS.

Part of the product identification is listing the proper or common chemical names of the hazardous ingredients. If the manufacturer chooses to claim the protection of a Trade Secret, then due care must be taken to ensure that all state regulations are followed which deal with trade secrets. It is important to provide adequate information while still protecting the manufacturer.

When the hazardous ingredients are listed, it is prudent to list them all. The composition does not need to be listed; this information is on the MSDS. The ingredients are listed, however, in order of decreasing concentration in the product. Ingredients below 1% concentration need not be listed, unless they are carcinogens, for which the cutoff is 0.1%. These latter guidelines are based on the OSHA requirements for MSDSs, not on a label standard.

Signal Words

Three signal words are defined to provide immediate recognition of the overall risk of the product. The signal words, in order of decreasing risk and their associated overall meaning are:

SIGNAL WORD	GENERAL MEANING
DANGER!	If this product gets in you or on you, immediate harm will be caused.
WARNING!	If this product gets in you or on you, in sufficient quantity, you will suffer harm.
CAUTION!	If this product gets in you or on you in large quantity over an extended time, you may be harmed.
The following term is not a SIGNAL Word, but does appear on a label to provide specific information.	
POISON! 	This product will make you ill if it enters your body by any route of entry.

The selection of the SIGNAL WORD is based on the most serious overall risk of using the product. The criteria for selecting the correct term are well-developed by the American National Standards Institute.[22] Tables 5 and 6 clearly state these criteria.

[22] This material is reproduced with permission from American National Standard for Hazardous Industrial Chemicals Precautionary Labeling Z129.1-1988, copyright 1988 by the American National Standards Institute; copies of this standard may be purchased from the American National Standards Institute at 11 West 42nd Street, New York, NY 10036.

TABLE 5

SELECTION OF LABEL LANGUAGE AND WORDING
Immediate Hazards

HAZARD	SIGNAL WORD	STATEMENT OF HAZARD	TARGET ORGAN(S)	PRECAUTIONARY MEASURES	INSTRUCTIONS IN CASE OF EXPOSURE
Highly Toxic (by ingestion)	DANGER!	MAY BE FATAL IF SWALLOWED	CAN CAUSE specify the organ(s) DAMAGE	Wash thoroughly after handling.	POISON! Get medical attention immediately. FIRST AID If swallowed (do not) induce vomiting immediately or as direct by a physician. Never give anything by mouth to an unconscious person.
Toxic (by ingestion)	WARNING!	HARMFUL IF SWALLOWED	CAN CAUSE specify the organ(s) DAMAGE	Wash thoroughly after handling.	FIRST AID: @ If swallowed, induce vomiting immediately as directed by medical personnel. Never give anything by mouth to an unconscious person. (Get medical attention (immediately)*) (Call a physician (Contact a poison control center)*)°
Highly Toxic (by absorption)	DANGER!	MAY BE FATAL IF ABSORBED THROUGH SKIN	CAN CAUSE specify the organ(s) DAMAGE	Do not get in eyes, on skin, or on clothing. Wash thoroughly after handling.	(Get medical attention (immediately*) (Call a physician (Contact a poison control center)*)°

Toxic (by absorption)	WARNING!	HARMFUL IF ABSORBED THROUGH SKIN	CAN CAUSE specify the organ(s) DAMAGE	Avoid contact with eyes, skin, and clothing. Wash thoroughly after handling.	FIRST AID: In case of contact, immediately flush eyes or skin with plenty of water for at least 15 minutes while removing contaminated clothing and shoes. (Get medical attention (immediately)*) (Call a physician (Contact a poison control center)*)° Wash clothing before reuse. (Destroy contaminated shoes.) + (thoroughly clean shoes before reuse.) +
Highly Toxic (by inhalation)	DANGER!	MAY BE FATAL IF INHALED	CAN CAUSE specify the organ(s) DAMAGE	Do not breathe (dust, vapor, mist, gas)*. Keep container closed. Use only with adequate ventilation.	POISON (Get medical attention (immediately)*) (Call a physician (Contact a poison control center)*)° FIRST AID: If inhaled, remove to fresh air. If not breathing, give artificial respiration. If breathing is difficult, give oxygen. (Get medical attention (immediately)*) (Call a physician)°
Toxic (by inhalation)	WARNING!	HARMFUL IF INHALED	CAN CAUSE specify the organ(s) DAMAGE	Avoid breathing (dust, vapor, mist, gas)*. Keep container closed. Use with adequate ventilation.	FIRST AID: If inhaled, remove to fresh air. If not breathing, give artificial respiration. If breathing is difficult, give oxygen. (Get medical attention (immediately)*) (Call a physician (Contact a poison control center)*)°

Table 5 (continued)

HAZARD	SIGNAL WORD	STATEMENT OF HAZARD	TARGET ORGAN(S)	PRECAUTIONARY MEASURES	INSTRUCTIONS IN CASE OF EXPOSURE
Strong Sensitizer, Lungs	(WARNING!) (DANGER!)°	MAY CAUSE (SEVERE)+ ALLERGIC RESPIRATORY REACTION	CAN CAUSE specify the organ(s) DAMAGE	Do not breathe (dust, vapor, mist, gas)*. Keep container closed. Use only with adequate ventilation.	FIRST AID: If inhaled, remove to fresh air. If not breathing, give artificial respiration. If breathing is difficult, give oxygen. (Get medical attention (immediately)*) (Call a physician)°
Irritant, Severe Respiratory	WARNING!	CAUSES RESPIRATORY TRACT IRRITATION (or specific site, if appropriate)	CAN CAUSE specify the organ(s) DAMAGE	Avoid breathing (dust, vapor, mist, gas)*. Keep container closed. Use with adequate ventilation.	FIRST AID: If inhaled, remove to fresh air. If not breathing, give artificial respiration. If breathing is difficult, give oxygen. (Get medical attention (immediately)*) (Call a physician)°
Irritant, Moderate Respiratory	CAUTION!	MAY CAUSE RESPIRATORY IRRITATION (or specific site, if appropriate)	CAN CAUSE specify the organ(s) DAMAGE	Avoid breathing (dust, vapor, mist, gas)*. Keep container closed. Use with adequate ventilation.	FIRST AID: If inhaled, remove to fresh air. If not breathing, give artificial respiration. If breathing is difficult, give oxygen. (Get medical attention (immediately)*) (Call a physician)°
Physiologically Inert Vapor or Gas	CAUTION!	(VAPOR) (GAS)* REDUCES OXYGEN AVAILABLE FOR BREATHING	CAN CAUSE specify the organ(s) DAMAGE	Keep container closed. Use with adequate ventilation. Do not enter (storage areas, confined spaces)* unless adequately ventilated.	FIRST AID: If inhaled, remove to fresh air. If not breathing, give artificial respiration. If breathing is difficult, give oxygen. (Get medical attention (immediately)*) (Call a physician)°

Irritant, Severe Eye	WARNING!	CAUSES EYE IRRITATION	CAN CAUSE specify the organ(s) DAMAGE	Avoid contact with eyes. Wash thoroughly after handling.	FIRST AID: In case of contact, immediately flush eyes with plenty of water for at least 15 minutes. (Get medical attention (immediately)*) (Call a physician)° Remove contaminating material from skin and clothing.
Irritant, Moderate Eye	CAUTION!	MAY CAUSE EYE IRRITATION	CAN CAUSE specify the organ(s) DAMAGE	Avoid contact with eyes. Wash thoroughly after handling.	FIRST AID: In case of contact, immediately flush eyes with plenty of water. (Get medical attention (immediately)*) (Call a physician)° if irritation persists.
Irritant, Severe Skin	WARNING!	CAUSES SKIN IRRITATION	CAN CAUSE specify the organ(s) DAMAGE	Avoid contact with eyes. Wash thoroughly after handling.	FIRST AID: In case of contact, immediately flush skin with plenty of water. Remove contaminated clothing. (Get medical attention (immediately)*) (Call a physician)° Wash clothing before reuse.
Irritant, Moderate Skin	CAUTION!	MAY CAUSE SKIN IRRITATION	CAN CAUSE specify the organ(s) DAMAGE	Avoid contact with skin and clothing. Wash thoroughly after handling.	FIRST AID: In case of contact, flush skin with plenty of water. Remove contaminated clothing. (Get medical attention (immediately)*) (Call a physician)° if irritation persists. Wash clothing before reuse.
Strong Sensitize, Skin	Warning!	MAY CAUSE ALLERGIC SKIN REACTION	CAN CAUSE specify the organ(s) DAMAGE	Avoid prolonged or repeated contact with skin. Wash thoroughly after handling.	FIRST AID: In case of contact, immediately wash skin with soap and plenty of water. Remove contaminated clothing. (Get medical attention (immediately)*) (Call a physician)° if symptoms occur. Wash clothing before reuse.

Table 5 (continued)

HAZARD	SIGNAL WORD	STATEMENT OF HAZARD	TARGET ORGAN(S)	PRECAUTIONARY MEASURES	INSTRUCTIONS IN CASE OF EXPOSURE
Corrosive, Eye	DANGER!	CAUSE (SEVERE)+ EYE BURNS	CAN CAUSE specify the organ(s) DAMAGE	Do not get in eyes. Avoid breathing (dust, vapor, mist, gas)*. Keep container closed. Use with adequate ventilation. Wash thoroughly after handling.	FIRST AID: In case of contact, immediately flush eyes with plenty of water for at least 15 minutes. (Get medical attention (immediately)*) (call a physician)°
Corrosive, Skin	DANGER!	CAUSES (SEVERE)+ SKIN BURNS	CAN CAUSE specify the organ(s) DAMAGE	Do not get on skin or clothing. Avoid breathing (dust, vapor, mist, gas)*. Keep container closed. Use with adequate ventilation. Wash thoroughly after handling.	FIRST AID: In case of contact, immediately flush skin with plenty of water for at least 15 minutes while removing contaminated clothing and shoes. (Get medical attention (immediately)*) (Call a physician)° Wash clothing before reuse. (Destroy contaminated shoes.)+ (Thoroughly clean shoes before reuse.)+
Corrosive, Other Target Organs	DANGER!	CAUSES (SEVERE)+ BURNS	CAN CAUSE specify the organ(s) DAMAGE	Do not get in eyes, on skin or on clothing. Do not breathe (dust, vapor, mist, gas)*. Keep container closed. Use with adequate ventilation. Wash thoroughly after handling.	FIRST AID: In case of contact, immediately flush eyes and skin with plenty of water for at least 15 minutes while removing contaminated clothing and shoes. (Get medical attention (immediately)*) (Call a physician)° Wash clothing before reuse. (Destroy contaminated shoes.)+ (Thoroughly clean shoes before reuse.)+

Table 5 (continued)

HAZARD	SIGNAL WORD	STATEMENT OF HAZARD	TARGET ORGAN(S)	PRECAUTIONARY MEASURES	INSTRUCTIONS IN CASE OF EXPOSURE
Extremely Flammable Liquid	DANGER!	EXTREMELY FLAMMABLE LIQUID AND VAPOR VAPOR MAY CAUSE FLASH FIRE	CAN CAUSE specify the organ(s) DAMAGE	Keep away from heat, sparks, and flame. Keep container closed. Use with adequate ventilation.	(See Table 3 for selection of appropriate fire-extinguishing statement.)
Flammable Gas	DANGER!	FLAMMABLE GAS MAY CAUSE FLASH FIRE	CAN CAUSE specify the organ(s) DAMAGE	Keep away from heat, sparks, and flame. Keep container closed. Use with adequate ventilation.	
Flammable Liquid	WARNING!	FLAMMABLE LIQUID AND VAPOR	CAN CAUSE specify the organ(s) DAMAGE	Keep away from heat, sparks, and flame. Keep container closed. Use with adequate ventilation.	
Flammable Solid	WARNING!	FLAMMABLE SOLID	CAN CAUSE specify the organ(s) DAMAGE	Keep away from heat, sparks and flame.	
Combustible Liquid	CAUTION!	COMBUSTIBLE LIQUID AND VAPOR	CAN CAUSE specify the organ(s) DAMAGE	Keep away from heat and flame.	
Pyrophoric Chemical	DANGER!	EXTREMELY FLAMMABLE CATCHES FIRE IF EXPOSED TO AIR	CAN CAUSE specify the organ(s) DAMAGE	Keep away from heat, sparks, and flame. Keep container tightly closed.	
Strong Oxidizer	DANGER!	STRONG OXIDIZER CONTACT WITH OTHER MATERIAL MAY CAUSE FIRE	CAN CAUSE specify the organ(s) DAMAGE	Keep from contact with clothing and other combustible materials. Do not store near combustible materials. Store in tightly closed container.	Remove and wash contaminated clothing promptly.

NOTES:

@ Do not use statement when vomiting is not appropriate.
* Select applicable word or words.
+ Use word or phrases when appropriate
o Select one.

TABLE 6
SELECTION OF LABEL LANGUAGE and WORDING
Delayed Hazards

HAZARD	SIGNAL WORD
CARCINOGENICITY	
Category I - Proven human carcinogenic substance	CANCER HAZARD; CONTAINS *fill in substance* , WHICH CAN CAUSE CANCER Risk of cancer depends upon duration and level of exposure.
Category II - Suspected human carcinogenic substance	SUSPECTED CANCER HAZARD; CONTAINS *fill in substance* , WHICH CAN CAUSE CANCER Risk of cancer depends on duration and level of exposure.
Category III - Proven animal carcinogenic substance of potential relevance to humans	POSSIBLE CANCER HAZARD; CONTAINS *fill in substance* , WHICH MAY CAUSE CANCER Risk of cancer depends upon duration and level of exposure.
TETRATOGENICITY	
Category I - Substances known to be teratogenic to humans	BIRTH DEFECT HAZARD; CONTAINS *fill in substance* , WHICH CAN CAUSE BIRTH DEFECTS
Category II - Substances that should be regarded as if they are teratogenic to humans	POSSIBLE BIRTH DEFECT HAZARD; CONTAINS *fill in substance* , WHICH MAY CAUSE BIRTH DEFECTS BASED ON ANIMAL DATA

The decision of using the word POISON!, and the associated Skull and Crossbones pictogram is rather straightforward. If the LD_{50} or the LC_{50} is less than 50 mg/kg, the word POISON! must be used in conjunction with the SIGNAL WORD: DANGER! In addition, if the Consumer Product Safety Commission requires the use of this pictogram, it must be used.

Determination of the Major Hazards

The initial analysis of the product should define those acute or chronic hazards for which warning must be given. The task is then to select succinct, pointed statements, which provide an immediate warning of these hazards. Examples of such statements are given in Tables 5 and 6. For example, if a product's major hazard is extreme flammability, then the following information may be selected:

DANGER!

EXTREMELY FLAMMABLE LIQUID AND VAPOR

VAPOR MAY CAUSE FLASH FIRE

If the product has several hazards of comparative risk, then each must be addressed in this section.

Target Organs

OSHA's policy on target organ information on labels has been affected by litigation. At the time of publication of this book, the courts have again reversed their positions (see footnote 6) and now agree with OSHA that labels must contain target organ information.

The target organ information often is obvious - corrosive chemicals attack the skin, eyes, mucous membranes, and respiratory tract. Just as frequently, however, such information may be obscure and difficult to obtain. The label preparer may need to examine the primary and derived literature to ensure completeness of this information.

Statement of Delayed Hazards

Along with information of acute effects of exposure, high risk chronic effects must be mentioned on the label. Thus, a product that contains an OSHA Known Human Carcinogen must have an appropriate warning statement on the label. For example,

DANGER!
CANCER HAZARD
CONTAINS *BENZENE* ,WHICH CAN CAUSE CANCER

Risk of cancer depends on duration and level of exposure.

If the chronic health affect is only expressed in animal studies or if the human data are insufficient to make a specific statement, the information must appear on the MSDS but need not appear on the label. The label preparer must use judgment in this.

Selection of Precautionary Measures Statements

For each acute hazard and each chronic hazard addressed on the label, the label preparer must include specific statements that explain methods for avoiding the hazard. Thus, the statement *"Keep away from heat, sparks, and flame. Keep container tightly closed."* For chemicals with special hazards, such as an organic peroxide with a Self-Accelerating Decomposition Temperature below room temperature, a statement such as *"Keep refrigerated. Do not allow to warm over 5°C,"* may be appropriate.

It is important that the Label Preparer only address hazards which are included in the "Statement of Immediate Hazards" or "Statement of Delayed Hazards" section of the label. Inclusion of methods for avoiding the hazard which extend those limits will tend to confuse the chemical user and may open the label up to litigation.

MSDS Availability

Neither OSHA nor ANSI require any mention of the MSDS on the label. Since the label and the MSDS are designed to work together to provide complete information to the chemical user, it is prudent to include a statement such as *"Refer to the MSDS for more information"* on the label.

Statements Related to Accidental Exposure

First-Aid Statements and Antidotes and Notes to Physicians

The actions to be taken in the event a person is accidently contaminated are well-established. Immediate removal of the material from the skin by a drench shower must be stated on the label. If inhalation or ingestion is a reasonable risk, then statements addressing these must be included. A typical statement to include on a label of a skin corrosive chemical is:

> *"In case of contact, immediately flush skin with plenty of water for at least 15 minutes while removing contaminated clothing and shoes. Get medical attention immediately."*

These statements are clearly of a "first-aid" nature. Yet ANSI includes a separate "First-Aid Statements and Antidotes" line item for its recommended label. This allows the label preparer to provide chemical specific first-aid information, as appropriate. For example, a statement related to the emergency use of a calcium gluconate gel might be appropriate on the label of a product that contains hydrofluoric acid.

The label preparer must select sufficient instructional and first-aid statements to allow adequate immediate care by a lay rescuer. The procedures recommended must not involve methods or materials that are not commonly available in locations where the chemicals are normally used.

Only a few chemicals have specific medical intervention procedures associated with acute exposure such that the industry practice is to provide that information as a "Note to Physician" on the label. A statement related to the use of amyl nitrite and sodium thiosulfate to counteract a cyanide exposure, for example, usually appears on the label of products containing a high concentration of cyanide. The label preparer must be aware of these specific situations and must be certain to include appropriate statements on the label.

Selection of Statements Related to Fire and Spills or Leaks

Complete instructions for response to a fire, spill, or leak cannot be included in the limited space of a precautionary label. Thus, the label preparer must develop a concise statement that addresses this requirement clearly and that enjoins the reader to seek additional information. The following statement has been used successfully for this purpose on many chemical labels:

> *"Only appropriately trained personnel should respond to chemical fires or spills. Use proper response procedures and maintain appropriate level of fire protection."*

Again, the judgment of the label preparer is essential for selecting the correct phrases to provide adequate warning.

Selection of Statements Related to Container Handling and Storage

By definition, precautionary labels will be placed on containers. Thus, ANSI recommends that some statements related to the handling and storage of the containers be included. The selection of the appropriate statements depends very much on the specific hazards of the chemical and the use to which the chemical will be applied.

The following statements might be appropriate on the label of an extremely flammable chemical drum:

> *"Use labeled flammable solvent container. Avoid contact*

with oxidizers and any other incompatible materials. Electrically ground container where necessary."

Other statements might be added, depending upon the circumstances.

Selection of Additional Useful Information

Statements such as "For Industrial Use Only" or "Keep Out of the Reach of Children" should be placed on containers as appropriate. The Consumer Product Safety Commission requires the latter statement on most hazardous chemical product labels. The former statement should be used for products that could conceivably be used by consumers, yet should only be used for industrial purposes. Chemicals related to the creative arts frequently are in this category.

ANSI recommends that the net contents of the container appear on the label. This information is generally not of a precautionary nature but is usually included.

Name and Address of Company

OSHA requires that the name and address (including phone number) of the chemical manufacturer, importer, or distributor appear on the label. This information can appear anywhere on the label. It need not appear on the precautionary information panel.

LABEL FORMAT AND STYLE

Chapter 5

INTRODUCTION

Once the label preparer has assembled the precautionary information, the label must be laidout. The dual purposes of the label must now be considered: product recognition and precautionary warning. This requires the label preparer to consider both format and style. Figure 15 illustrates various styles and formats of labels.

FORMAT

Three basic formats for the label are in general use. These are illustrated in Figure 14, 15, and 16. The general formats are to (1) (Figure 14) place the product identification information to the side of the precautionary information; (2) (Figure 15) place the product identification information in a stylized manner relative to the precautionary information; and (3) (Figure 16) place the precautionary information on a separate panel, usually located on the opposite side of the container. Care must be taken with the latter format to ensure that the precautionary information is prominent. No one format has any particular advantages; the decision is usually one of corporate identity.

The label format is very much involved with product identity. A chemical manufacturer will generally use the same format for all of its container labels. Compare, for example, Figures 2, 3, and 15. These are all Eastman Fine Chemicals produced by the Eastman Kodak Company. All three labels, and indeed, all of the Kodak Fine Chemicals, use exactly the same label format, thus enhancing product recognition. Figure 16 is used by Cohler Enterprises, Inc. on containers of gold plating solution, which contains cyanide salts.

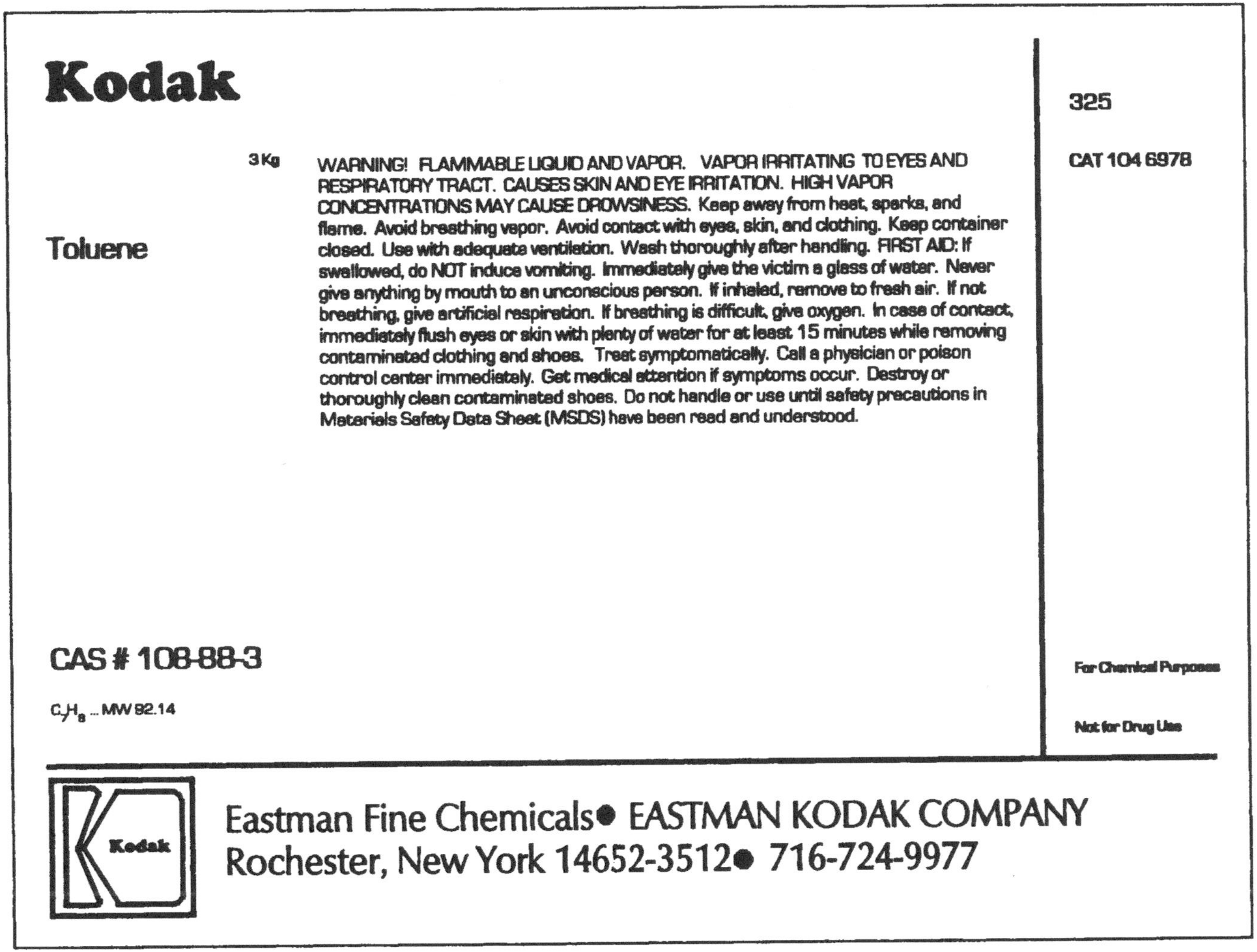

Figure 14 Label illustrating a stylized format in which chemical identity is separate from precautionary statements.

DIRECTIONS FOR USE: It is a violation of Federal law to use this product in a manner inconsistent with its labeling.

DEXOL PREDATOR CARPENTER ANT KILLER DUST is a ready-to-use insecticide dust which has given effective knockdown and residual control of the pests listed. DEXOL PREDATOR CARPENTER ANT KILLER DUST is for residual pest control in and on buildings and structures and their immediate surroundings including houses, garages, trailers, campers, storage beds, and apartment buildings.

INDOOR USE: DEXOL PREDATOR CARPENTER ANT DUST is intended for application with a shaker or puffer container to hiding and runway areas and those places where pests are found. The amount to be applied will vary with the site, but should usually be in the range of 1/3 to 2/3 oz. of DEXOL PREDATOR CARPENTER ANT DUST per sq. yd.

CARPENTER ANTS: Apply around doors and window and other places where ants enter premises and where they crawl and hide. Also force dust into infested wood through existing openings or newly drilled small holes. When possible locate and treat nest and surrounding areas.

ANTS, CRICKETS, SILVERFISH AND FIREBRATS: Thoroughly dust all areas where these pests crawl and hide, especially cracks and crevices around sinks, stoves, refrigerators and cabinets; behind baseboards; around doors and windows and in attics, basements and crawl spaces. For ants, apply to ant trails and around doors and windows where they may enter premises.

SPIDERS: Apply along and behind baseboards, to window and other door frames, corners, pipes, storage localities, attics, crawl spaces and other areas over which these pests may crawl.

STORAGE & DISPOSAL

Storage: Keep pesticide in original container. Avoid contamination of feed and foodstuffs. Store in a cool, dry place, preferably in a locked storage area. **Pesticide Disposal:** Securely wrap original container in several layers of newspaper and discard in trash. **Container Disposal:** Do not reuse canister. Discard canister in trash.

PRECAUTIONARY STATEMENTS
HAZARDS TO HUMANS AND DOMESTIC ANIMALS
CAUTION

Harmful if swallowed, inhaled or absorbed through the skin. Do not breathe dust. Do not allow to contact skin, eyes or clothing. Wash hands and exposed skin before eating, drinking or smoking and after handling. Wash all contaminated clothing thoroughly before reuse.

PRECAUTIONS

Do not use on or contaminate fruit, vegetables, or other food or feed crops. Do not apply to humans, animal clothing or bedding. Do not contaminate feed or food products or food preparations surfaces, dishes, kitchen utensils and food containers.

STATEMENT OF PRACTICAL TREATMENT

If Swallowed: Drink 1 or 2 glasses of water and induce vomiting by touching back of throat with finger. Do not induce vomiting or give anything by mouth to an unconscious person. Get medical attention. **If Inhaled:** Keep patient strictly at rest and transfer to hospital immediately to get medical attention. **If On Skin:** Wash with soap and warm water. **If In Eyes:** Rinse thoroughly with clean water. NOTE TO PHYSICIAN: Contains anticholinesterase carbamate insecticide. In cases of overexposure, atropine sulfate-2 mg. (equivalent to 1/30 grain adult dose) should be given, preferably by injection, and repeated as necessary until fully atropinized.

ENVIRONMENTAL HAZARDS

This pesticide is toxic to fish and wildlife. Do not apply directly to water or wetlands (swamps, bogs, marshes and potholes). Runoff from treated areas may be hazardous to aquatic organisms in neighboring areas. Do not contaminate water when disposing of equipment wash waters. This product is highly toxic to bees exposed to direct treatment on blooming crops or weeds. Do not apply this product or allow it to drift to blooming crops or weeds if bees are visiting the treatment area.

NOTICE: Seller makes no warranty, expressed or implied, concerning the use of this product other than indicated on the label. Buyer assumes all risk of use and/or handling of this material when such use and/or handling is contrary to label instructions.

QUESTIONS: CALL (800) 421-2934

DEXOL INDUSTRIES
Torrance, CA 90501 Rev. 3/91 **EPA Reg. No. 192-148**
EPA Est. No. 192-CA-1

UPC
POSITION

DEXOL

PREDATOR®
CARPENTER
ANT KILLER DUST

Kills Carpenter Ants Fast

ALSO CONTROLS:
ANTS, CRICKETS, SILVERFISH, FIREBRATS AND SPIDERS

ACTIVE INGREDIENTS:
Bendiocarb: (2,2-dimethyl-1,3 benzodioxol-4-yl methylcarbamate*) 1.0%
INERT INGREDIENTS: 99.0%
TOTAL 100.0%

*Protected by U.S. Patent No. 3,736,338

Keep Out of Reach of Children
CAUTION
See back panel for additional precautionary statements

NET WT. 12 OZ.

Figure 15 Label illustrating side-by-side format.

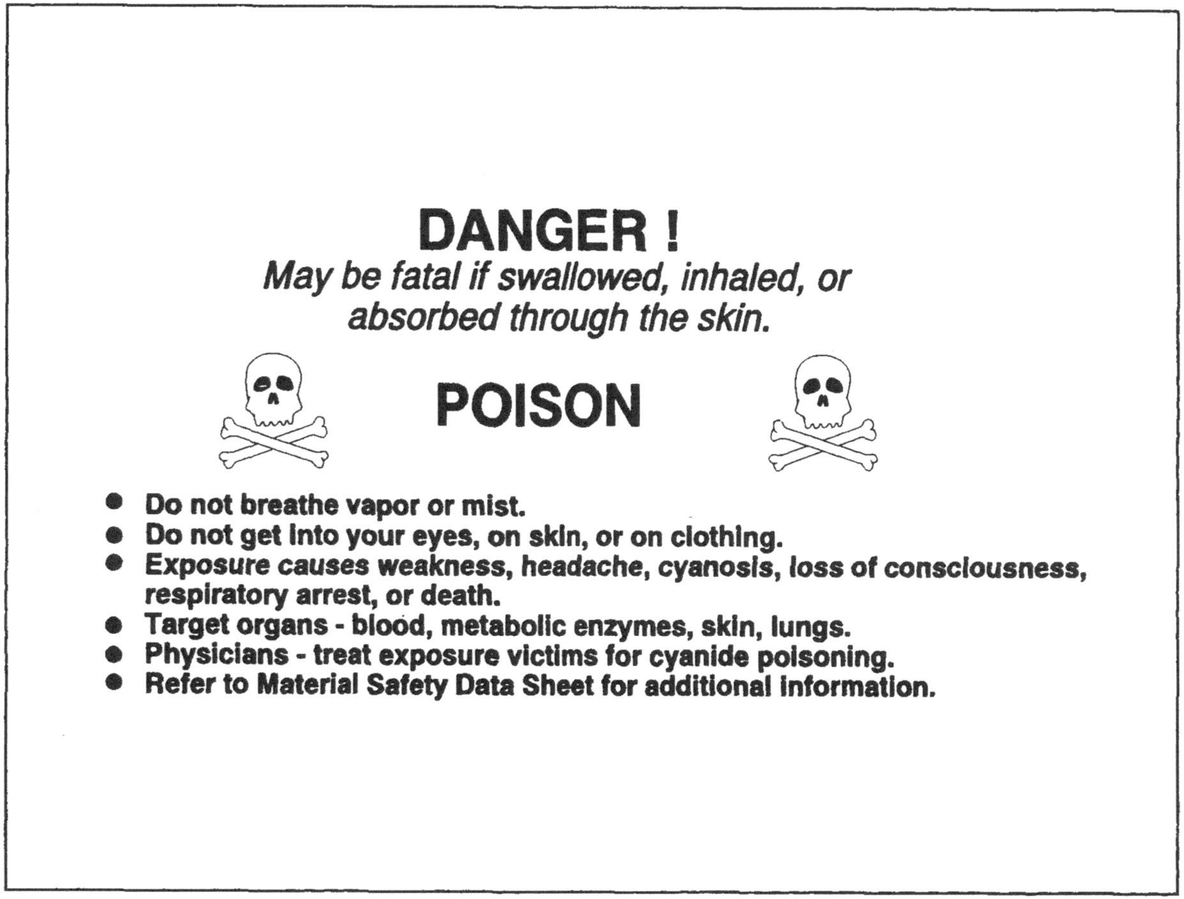

Figure 16 Precautionary label for a cyanide-containing gold plating solution. This label is placed on the back of the container, separate from the principal display panel.

Style

The style elements that the label preparer must consider include:

CHARACTER
LEGIBILITY
INFORMATION
READABILITY
VISIBILITY
COLOR
COMMUNICATIONS

These elements[23] must be considered to design a label that effectively communicates the hazard information within the available space for the label.

Character Legibility

The basic element that comprises the legibility of the individual characters is the font. This includes the character stroke-width and the width-to-height ratio. These comprise the basic shape of the characters. Many references[24] exist to assist in the selection of these elements. These should be consulted as needed by the label preparer.

Clean fonts, such as Univers, Helvetica, or CG Times generally have acceptable legibility for precautionary label use. The point size used depends markedly on the overall size of the label. The product identification is usually large, 24 - 36 point, for example. The Signal Word should be 12 - 14 point, Bold and the statement of hazards 1 - 2 points smaller. The details of the precautionary information are generally not smaller than 8 point.

---

23 A more detailed discussion of this topic may be found in "Design of Warning Labels and Instructions" by Joseph P. Ryan, Van Nostrand Reinhold, New York, 1991

24 American National Standard Institute A13.1-1981 "The Scheme for Identification of Piping Systems"
U.S. Department of Defense, 1975 MIL Spec. MIL-HBK-759
U.S. Department of Defense, 1981 MIL Spec. MIL-STD-1472C

A number of issues must be evaluated related to legibility. Some, such as viewing light intensity may not be applicable in the industrial setting. The label preparer must evaluate this on a case-by-case basis.

Readability

The ability of the viewer to recognize the information content of the text is referred to as "readability". Again, the selection of the font is critical to the ability of the viewer to discern the information. Clean fonts, of medium weight are best. Illustrated below are three fonts, in 12, 10, and 8 point. The 10-point presentation is bold.

Univers - 12 point **This line, in 10 point, is bold** This line is presented in 8 point
Helvetica - 12 point **This line, in 10 point, is bold** This line is presented in 8 point
CG - Times - 12 point **This line, in 10 point, is bold** This line is presented in 8 point *This line is presented in 8 point italics*

Fancy fonts, including *italics* should be avoided on precautionary labels. Upper-case type is harder to read than lower-case type. Only the first letter of a sentence and of proper names should be in upper-case. If an occasional word is placed in upper-case, the reason must be for emphasis, and the word must be carefully chosen.

Other characteristics that lead to loss of readability and that should be avoided included hyphenated words, abbreviations, and centered text.

Visibility

In order to be effective, precautionary labels must be easily read from a safe distance. Routine use of labels places them on containers. For routine use, the safe distance to a chemical container is quite close; often the container is in the hands of the chemical user. Thus, the visibility considerations for a chemical container precautionary label is less than 1 meter.

This distance limitation allows considerable relaxation of some of the design considerations, since the critical precautionary elements need not be extra large to achieve visibility. The viewing angle and the time the chemical user has to view the precautionary information are not particularly limited. The important issue is to place the chemical identification and the signal word within sufficient proximity to allow for the recognition of both within the initial view of the label.

Color

The use of color on a precautionary label can be quite effective. Improper use of color can be misleading or can be the direct cause of accidents.

Red is reserved for flammable materials. Yellow is reserved for oxidizers. Misuse of these colors could lead to the accidental mixing of these incompatible chemicals resulting in a fire or explosion. The label preparer must be sensitive to these type issues to avoid issues related to product recognition or advertising from confusing precautionary warning information.

Color is extremely useful in conveying precautionary information. The most common color scheme is that adopted by the U.S. Department of Transportation. Insofar as the label preparer is able, the color scheme in Table 7 should be used.

TABLE 7 DOT Labels

PRIMARY HAZARD	COLOR(S)	PICTOGRAM
Explosives and Blasting Agents	Black letters on an orange background	
Flammable Gases	White letters on a red background	
NonFlammable Gases	White letters on a green background	

PRIMARY HAZARD	COLOR(S)	PICTOGRAM
Poison Gases	Black letters on a white background	
Oxygen (oxidizing gases)	Black letters on a yellow background	
Flammable and Combustible Liquids	White letters on a red background	

PRIMARY HAZARD	COLOR(S)	PICTOGRAM
Flammable Solids	Black letters on a red and white vertical stripe background	
Spontaneously Combustible	Black letters on a red and white field, split horizontally, with the red half on the bottom	
Dangerous When Wet	White letters on a dark blue background	

PRIMARY HAZARD	COLOR(S)	PICTOGRAM
Oxidizers	Black letters on a yellow background	
Poisonous solids and liquids	Black letters on a white background	
Radioactive materials	Black letters on a yellow and white field, split horizontally, with the yellow half on the bottom	

PRIMARY HAZARD	COLOR(S)	PICTOGRAM
Corrosive materials	White letters on a white and black field, split horizontally, with the black field on the bottom	
Biohazard and etiologic agents	Black letters and characters on an orange field	

Use of these colors will provide the chemical user with a recognition matrix that is consistent with the commonly accepted transportation colors. While other color schemes are used, such as that cited by OSHA in 29 CFR 1910.145, these are for warning signs, not for precautionary labels.

Pictograms

Pictograms convey more information than any other label element. Table 7 includes the most recognized symbols - the DOT Hazard Class pictograms - for each hazard. Consistent use of the pictograms and the colors will lead to uniform recognition of potential hazards.

LABEL EVALUATION
Chapter 6

PRECAUTIONARY LABELS FOR CHEMICAL CONTAINERS

INTRODUCTION

A label preparer must have some systematic method for evaluating the information on a precautionary label. Using a systematic method removes as much of the subjectivity from the evaluation as possible. Evaluations must be performed on the preparer's work as well as on existing labels of the company. Presented in Tables 8 and 9 are evaluation forms designed to review label information quickly and effectively. These forms are not designed to be used by persons unfamiliar with label preparation. Significant judgment enters the evaluation process and must be recognized.

Label Evaluation Forms

The evaluation forms presented address both consumer product and commercial or industrial labels.

<u>Consumer Product Label Evaluation Forms</u>

TABLE 8
CONSUMER PRODUCT LABEL EVALUATION FORM

CONSUMER PRODUCT LABEL EVALUATION - 16 CFR 1500.121

PRODUCT NAME:
MANUFACTURER:

LABEL REQUIREMENTS	ADEQUATE	INADEQUATE	MISSING
Signal Word DANGER! WARNING! CAUTION!			
IS "POISON" REQUIRED? PRESENT?			
Statement of principal hazard			
The common or usual or chemical name of the hazardous substance(s)			
Name and place of business of manufacturer, packer, distributor, or seller			
Statement of precautionary measures for each hazard, if more than one hazard exists			
Instructions for special handling when appropriate for each hazard, if appropriate			
The statement "KEEP OUT OF REACH OF CHILDREN" or its equivalent			
First-aid instructions when appropriate for each hazard, if appropriate			

HAZARDOUS CONSTITUENTS OF THIS PRODUCT

1.	5.
2.	6.
3.	7.
4.	8.

COMMENTS

TABLE 8, continued

SPECIAL CONSIDERATIONS FOR CONSUMER LABELS:

NAMED SUBSTANCES IN THE FEDERAL CAUSTIC POISON ACT THAT REQUIRE THE WORD "POISON" INSTEAD OF ANY OTHER SIGNAL WORD.

Hydrochloric acid	>10%
Nitric acid	>5%
Oxalic acid	>10% or any salt
Acetic acid	>20%
Potassium hydroxide	>10%
Silver nitrate	>5%
Sulfuric acid	>10%
Carbolic acid (phenol)	>5%
Hypochlorous acid	>10%
Sodium hydroxide	>10%
Ammonia water	>5%

PRODUCTS WHICH REQUIRE SPECIAL LABELING

Diethylene glycol and mixtures containing 10% (w/w) or more
Ethylene glycol and mixtures containing 10% (w/w) or more

Benzene	>5%
Toluene	>10%
Xylene	>10%
Petroleum distillates	>10
Methyl alcohol	>4%
Turpentine	>10%

BANNED HAZARDOUS SUBSTANCES

MIXTURES THAT ARE INTENDED FOR APPLICATION TO INTERIOR FLOORS OR WALLS THAT ARE MARKED "EXTREMELY FLAMMABLE"

CARBON TETRACHLORIDE AND MIXTURES CONTAINING CARBON TETRACHLORIDE

LIQUID DRAIN CLEANERS CONTAINING 10%(w/w) or more SODIUM HYDROXIDE AND/OR POTASSIUM HYDROXIDE EXCEPT IF PROPERLY PACKAGED AS PER THE POISON PREVENTION PACKAGING ACT OF 1970.[25]

[25] Public Law 91-601, 84 Stat. 1670-74; 15 U.S.C. 1471-76

TABLE 9
INDUSTRIAL CHEMICAL PRODUCT LABEL EVALUATION FORM

INDUSTRIAL CHEMICAL PRODUCT LABEL EVALUATION –
ANSI Z129.1-1988 & 29 CFR 1910.1200

PRODUCT NAME:
MANUFACTURER:

LABEL REQUIREMENTS	ADEQUATE	INADEQUATE	MISSING
IDENTIFICATION OF THE CHEMICAL OR ITS HAZARDOUS COMPONENTS			
SIGNAL WORD DANGER! WARNING! CAUTION!			
STATEMENT OF HAZARD(S)			
PRECAUTIONARY MEASURES			
INSTRUCTIONS IN CASE OF CONTACT OR EXPOSURE			
ANTIDOTES			
NOTES TO PHYSICIANS			
INSTRUCTIONS IN CASE OF FIRE, SPILL, OR LEAK			
INSTRUCTIONS FOR CONTAINER HANDLING AND STORAGE			
MANUFACTURER'S NAME, BUSINESS ADDRESS, AND PHONE NUMBER			
TARGET ORGANS			
MSDS AVAILABILITY & LOCATION (OSHA suggested; not required)			
"POISON" & SKULL AND CROSSBONES Needed for highly toxic chemicals. This does not replace signal word. 			

HAZARDOUS CONSTITUENTS OF THIS PRODUCT

1.	5.
2.	6.
3.	7.
4.	8.

COMMENTS

SPECIAL PURPOSE LABELS
Chapter 7

PRECAUTIONARY LABELS FOR CHEMICAL CONTAINERS

ENVIRONMENTAL LABELS

Two major types of materials require environmental warning labels: hazardous wastes and pesticides. The label content and information for both is defined by the U.S. Environmental Protection Agency. The regulations defining label information and usage for hazardous wastes are issued under the Resource Conservation and Recovery Act (RCRA). The regulations defining label information and usage for pesticides are issued under the Federal Insecticide, Fungicide, and Rodenticide Act (FIFRA). The RCRA label regulations are published at 40 CFR 262.32. The pesticide regulations are published at 40 CFR 162 - 180.

Hazardous Waste Labels

Figure 17 EPA Hazardous Waste label, showing the additional California label requirements.

The EPA requires minimum information on a container of hazardous waste. The specific requirement states:

> *Before transporting hazardous waste or offering hazardous waste for transportation off-site, a generator must mark each container of 110 gallons or less used in such transportation with the following words and information displayed in accordance with the requirements of 49 CFR 172.304:*
>
> *HAZARDOUS WASTE - Federal Law Prohibits Improper Disposal. If found, contact the nearest police or public safety authority or the U.S. Environmental Protection Agency.*
>
> *Generator's Name and Address*
>
> ________________________________.
>
> *Manifest Document Number*
>
> ________________________________

Many states, however, require additional information on containers of hazardous waste. The additional information required by California is shown on the California Hazardous Waste label in Figure 17. The EPA has argued that other labels, in particular, DOT shipping labels will impart adequate warning information. The Hazard Communication Standard specifically excludes applicability to hazardous waste; thus no MSDS is required nor are standard chemical warning labels.

When completing a Hazardous Waste label, the generator, or other person attaching the label to the container must fully complete all required information. In particular, the **Accumulation Start Date** must be placed on the label and indicate the day the *first quantity of hazardous waste is placed in the container.* Failure to do this is one of the leading RCRA violations. Another leading RCRA violation is failure to transfer the hazardous waste to a permitted RCRA facility within 90 days of the accumulation start date.

Pesticide Label

No chemical regulated under FIFRA may be sold, or offered for sale in the United States until the U.S. EPA has reviewed and approved the warning label for the product. In addition, many states require a state review of the label. Following such

PRECAUTIONARY STATEMENTS

HAZARDS TO HUMANS & DOMESTIC ANIMALS

CAUTION: Harmful if swallowed, inhaled or absorbed through the skin. Avoid breathing vapors or spray mist. Avoid contact with skin, eyes and clothing. Wash thoroughly after handling and before eating or smoking. Avoid contamination of feed or foodstuffs. Do not permit children or pets to go on to treated area until the spray has dried completely. Do not contaminate ornamental fish ponds. Food utensils such as teaspoons or measuring cups should not be used for food purposes after contact with pesticides.

STATEMENT OF PRACTICAL TREATMENT

If Swallowed: Call a physician or Poison Control Center immediately. DO NOT INDUCE VOMITING UNLESS UNDER MEDICAL SUPERVISION. **If Inhaled:** Remove victim to fresh air. Apply artificial respiration if indicated. **If On Skin:** Remove contaminated clothing. Wash affected areas with soap and water. **If In Eyes:** Flush eyes with plenty of water. Get medical attention of irritation persists. NOTE TO PHYSICIAN: Chlorpyrifos is a cholinesterase inhibitor. Treat symptomatically. Atropine by injection is an antidote.

ENVIRONMENTAL HAZARDS

This product is toxic to fish, birds and other wildlife. Do not apply directly to water. Do not apply when weather conditions favor drift from treated areas. Do not contaminate water by cleaning of equipment or disposal of wastes. This pesticide is highly toxic to bees exposed to direct treatment or to residues remaining on the treated area. Protective information may be obtained from your Cooperative Agricultural Extension Service.

NOTICE: Buyer assumes all risk of use, storage, or handling of this material not in strict accordance with directions given herewith.

DEXOL INDUSTRIES — **EPA REG. No. 192-142**
Torrance, CA 90501 — **EPA Est. No. 192-CA-1**

Figure 18 Pesticide label which complies with the Federal FIFRA requirements and the state labeling requirements.

a review, the EPA (and the state agency, if applicable) issues a registration number, which must appear on the label. The chemical manufacturer may not alter the content, format, or usage of the label without express consent of the EPA (and state). Figure 18 presents an example of the product identification and precautionary statements of a pesticide label. The use information is included in a panel, which appears to the right of the product identification. As can be seen on this label, it carries an EPA registration number, as required by FIFRA.

Biohazard Label

A new class of regulated materials for both OSHA and EPA for which precautionary information is required is "biohazards." A biohazard is any agent, organism, or pathogen that may reasonably be expected to cause illness or disease in a human. Human blood, because of its potential to contain the Human Immunodeficiency Virus (HIV) or the Hepatitis B Virus (HBV), is considered a biohazard and is regulated by both OSHA as a biohazard and by DOT as an "etiological agent." Human blood that is a waste is regulated in several states as a medical waste or a hazardous waste.

OSHA and DOT require appropriate precautionary information be placed on containers that contain a biohazard. The label must incorporate the etiological agent symbol, Figure 19, and must be clearly visible on outside packaging. The new OSHA "Bloodborne Pathogen Standard" (29 CFR 1910.1030) defines the usage of the biohazard label. OSHA maintains that any container from which an exposure may be reasonably expected must be so labeled. The DOT requirements are found in 49 CFR 173.134.

Figure 19 Biohazard label.

Ozone Depleter Label

Effective on May 14, 1993, no container in which a Class I or Class II ozone depleter substance is stored or transported may be introduced into interstate commerce unless it bears a clearly legible and conspicuous label stating: *"WARNING: Contains [substance name], a substance which harms public health and environment by destroying the ozone in the upper atmosphere."*

EXAMPLES OF CLASS I and CLASS II OZONE DEPLETING COMPOUNDS	
CLASS I OZONE DEPLETING COMPOUNDS	CLASS II OZONE DEPLETING COMPOUNDS
Freon 11, 12, 13 Halons Carbon tetrachloride	All hydrochlorofluorocarbons (HCFCs)

Effective May 14, 1993, products manufactured with a Class II substance must be labeled if a substitute product or manufacturing process exists which:

(1) does not rely upon listed substances;

(2) reduces the overall risk to human health and the environment; and

(3) is currently available.

All products manufactured with Class I substances must be labeled unless the EPA affirmatively makes a determination that no substitute process or product exists that meets the above tests. A product requiring labeling may not be introduced into interstate commerce unless it bears a clearly legible and conspicuous label stating: *"Warning: Manufactured with: [substance name], a substance which harms public health and environment by destroying the ozone in the upper atmosphere."* This labeling requirement will apply to all products manufactured with Class I or Class II substances after January 1, 2015.

Marine Pollutant Label

On July 14, 1991, 19,000 gallons of a pesticide named METAM SODIUM was spilled into the Sacramento River, in northern California. The product hydrolyzes in water to release hydrogen sulfide. The spill contaminated approximately 50 miles of the river, killing most aquatic life. The resulting cloud of hydrogen sulfide forced the closing of all roadways in the area and the evacuation of all residents. The chemical was not regulated for shipment in the U.S. but was regulated as a Marine pollutant by the International Maritime Organization (IMO). As a result of that accident, the U.S. DOT issued a Final Rule published in the Federal Register on November 5, 1992, which incorporates the IMO Marine Pollutants into the regulatory framework of the DOT Hazardous Materials Shipping Regulations. The DOT accomplished this by adopting the IMO list of Marine Pollutants and developing a regulatory procedure for each modality - vessel, rail, highway, and air. The Final Rule[26] contains specific requirements for labeling both bulk and nonbulk containers that are being shipped by vessel. Some examples of chemicals regulated as marine pollutants include cyanide compounds, most mineral acids, and most pesticides.

Package Markings

The DOT has adopted the marking used by IMO for marine pollutants. This is shown in Figure 20. The marking must be in black letters on a white background, or in contrasting colors to the surface to which it is attached. For non-bulk packages the

26 This section could be included in the Chapter 2, but was placed in this location because of the unique nature of the precautionary information

mark must be at least 100 mm on a side, configured as a square on point. For bulk packaging, the mark must be at least 250 mm on a side, similarly configured.

Figure 20 DOT Marine Pollutant label.

Nutritional Label for Food

The Food and Drug Administration (FDA) has developed a new food labeling requirement that will be used on food products starting in the Spring of 1993. This new label requirement is an attempt to standardize the information provided on food labels, especially the method of percentage calculated for nutritional components. The FDA feels this labeling requirement will help consumers determine how the product fits into their diet and will provide consistent nutritional information.

This new label requirement, Figure 21, not only provides the Percentage of U.S. Recommended Daily Allowances (U.S. RDA) for vitamins and minerals; it also provides the Percent of a Daily Value for fat, cholesterol, sodium, carbohydrates and protein. The Percent of a Daily Value based on a 2,000 calorie diet, so the Daily Values will vary for different caloric needs. The requirement provides for a general guide for how much, or how little, of each of the key nutrients is needed based on a 2,000 and 2,500 calorie diet.

NUTRITION FACTS
Serving size 1/2 cup (114g)
Servings per container 4

Amount per serving

Calories 260 Calories from fat 120

	% Daily Value*
Total Fat 13g	**20%**
Saturated Fat 5g	**25%**
Cholesterol 30mg	**10%**
Sodium 660g	**28%**
Total Carbohydrate 31g	**11%**
Sugars 5g	
Dietary Fiber 0g	**0%**
Protein 5g	

Vitamin A 4% Vitamin C 2%
Calcium 15% Iron 4%

*Percent (%) of a Daily Value is based on a 2,000 calorie diet. Your Daily Values may vary higher or lower depending or your caloric needs:

Nutrient		2,000 Calories	2,500 Calories
Total Fat	Less than	65g	80g
Saturated Fat	Less than	20g	25g
Cholesterol	Less than	300mg	300mg
Total Carbohydrate		300g	375g
Fiber		25g	30g

1g Fat = 9 calories
1g Carbohydrate = 4 calories
1g Protein = 4 calories

Figure 21 Nutritional information label.

Child Drowning Warning

The California legislature enacted legislation which goes into effect on September 1, 1993, which requires a "Child Drowning Warning" on all 4 to 6 gallon open head plastic and metal pails. The label, which must be attached to the pail, is shown in Figure 22. As shown, it must be in both English and Spanish.

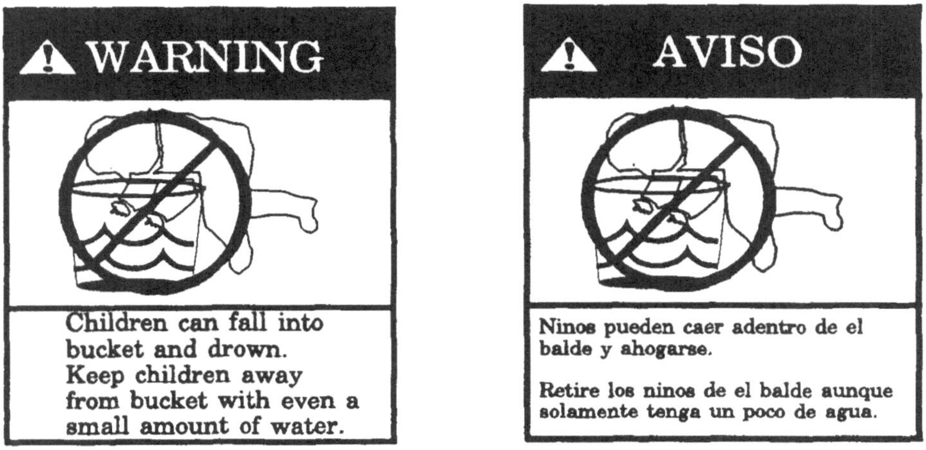

Figure 22 Child Drowning Warning label which must be attached to 4 - 6 gallon open head pails that are used in California after September 1, 1993.

USING LABELS

Chapter 8

PRECAUTIONARY LABELS FOR CHEMICAL CONTAINERS

INTRODUCTION

The ultimate purpose of a precautionary label is to provide information needed to the chemical user to prevent injury or illness. The chemical user must use this information effectively in order to achieve maximum benefit. The purpose of this chapter[27] is to provide the chemical user with the tools needed to work safely.

Case Histories[28]

A. *More than fifty employees were injured when chlorine gas was formed as a result of the delivery of the wrong chemical to a wastewater treatment plant. A bulk tanker was accidently loaded with sodium hypochlorite, rather than the ordered commodity – aluminum sulfate. The error was not detected prior to delivery, and the acid aluminum sulfate reacted with the sodium hypochlorite to release a large cloud of chlorine. Investigation of the accident revealed that the loading rack at the supplier allowed the delivery of five different chemicals via a set of five identical hoses controlled by five identical valves. Neither the hoses nor the valves were labeled.*

B. *An employee was severely burned as the result of the accidental mixing of solid citric acid with a solution of sodium hydroxide. The label on the burlap sack of citric acid was written in Kanjii characters, which the English-speaking victim was unable to read.*

C. *A construction worker was severely burned when he mixed an organic peroxide with a cobalt octanate solution. While the peroxide container was very well labeled, the container of cobalt solution was unidentified and had no precautionary information.*

---

27 The material in this chapter is based on Chemical Safety Associates' training manual, Hazardous Chemical Safety

28 The following case histories are taken from the files of Chemical Safety Associates, Inc. - the essential facts are presented unchanged; sufficient facts have been altered to protect the identities of all involved parties

Each of these cases occurred in the United States, at least five years *after* the implementation of the Hazard Communication Standard. In each case, multiple failures contributed to the accident, including, in some measure, actions of the victim(s). Inadequate chemical labeling was, however, a major proximate factor in every case.

The first time a chemical user handles a chemical, and, whenever a significant period has passed since the information has been reviewed, chemical users **MUST READ THE DETAILED PRECAUTIONARY INFORMATION** on the label. This information states the following:

WHAT IS THE HAZARD
HOW TO AVOID THE HAZARD
HOW TO RECOGNIZE EXPOSURE
WHAT TO DO IN CASE OF EXPOSURE

Information may also be included concerning proper handling, storage, and shipment of the chemical. Most companies also include minimum information for spill response and disposal.

READ THE LABEL - IT IS THE KEY TO CHEMICAL SAFETY

RULES FOR USING LABELS

Supervisors constantly remind employees to "Read the Label;" but seldom tell them what this injunction means. Clearly, it is unreasonable to assume that a person will read every word on a chemical label each time they use a chemical. The worker should, however, **IDENTIFY THE PRODUCT** and **DETERMINE THE SIGNAL WORD** each time they go to use a chemical. If they do not know why the particular signal word is associated with the product, then the worker should review the precautionary information; if they still have questions, they should refer to the MSDS, and finally to their supervisor.

Because much chemical usage is repetitive and because people become careless doing repetitive tasks, protective habits must be developed to compensate for human fallibility. The primary habit related to product identification is to **IDENTIFY THE PRODUCT AT LEAST THREE TIMES -**

ONCE BEFORE REMOVING THE CONTAINER FROM STORAGE;
ONCE BEFORE OPENING THE CONTAINER; and,
ONCE BEFORE INTRODUCING THE PRODUCT INTO THE PROCESS.

Experience clearly indicates that this procedure will prevent many mishaps by accidentally confusing chemicals.

These concepts have been summarized in the Rules of Chemical Labeling[29] presented in Table 10.

29 Hazardous Chemical Safety by Chemical Safety Associates, Inc., 1985

TABLE 10

Rules of Chemical Labeling

- ✓ ALL CONTAINERS, PIPES , PROCESS VESSELS, AND CHEMICAL STORAGE AREAS MUST BE LABELED.
- ✓ ALL CHEMICAL LABELS MUST IDENTIFY THE PRODUCT AND THE HAZARDS ASSOCIATED WITH ITS USE.
- ✓ ASSOCIATE THE "SIGNAL WORD" WITH THE PRODUCT BEFORE REMOVING FROM STORAGE.
- ✓ ANYONE USING A CHEMICAL MUST READ THE LABEL AT LEAST THREE TIMES BEFORE USING THE PRODUCT:
  - ★ IDENTIFY THE PRODUCT BEFORE REMOVING IT FROM STORAGE;
  - ★ IDENTIFY THE PRODUCT BEFORE OPENING THE CONTAINER;
  - ★ IDENTIFY THE PRODUCT BEFORE POURING IT INTO THE REACTION VESSEL.
- ✓ EVERY SUBDIVISION CONTAINER MUST BE LABELED PROPERLY.

All containers must be labeled, except those that will remain in the immediate control of the person who filled it, during the entire work period. If the container is left unattended, even during a rest break, it must be labeled. For manufacturing and process operations, this is easy - every container is labeled. The real problem is in laboratories, where very small beakers or sample tubes are used. Reasonable design, however, makes labeling the smallest container possible, however. If the container is so small that you cannot place a label on it, a 1 mL beaker for example, place it in a larger container, a 100 mL jar, with a lid, and label the outer container. Examples of in-house container labels were given in Figures 11 - 13 and 23 .

The information that should be on such a laboratory label, at a minimum should be **(1) the name of the chemical, (2) the SIGNAL WORD, (3) the name of the person responsible for the container, and (4) the date the container was filled**. Company policy or practice may require more information, such as laboratory notebook page number or primary hazard information. Preprinted in-house labels, such as that shown in Figure 23, help make laboratory labeling easy.

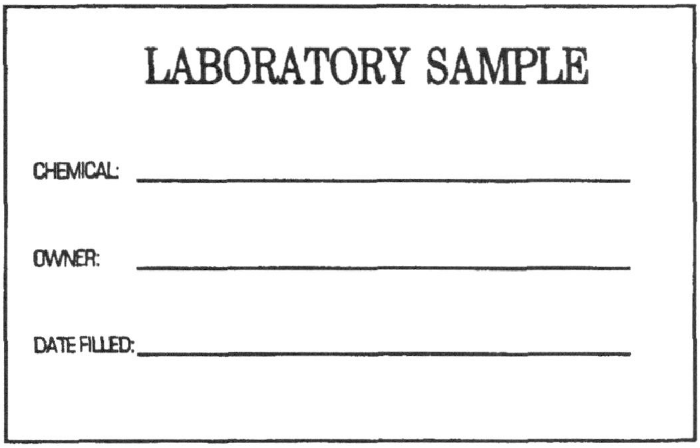

Figure 23 Laboratory label.

MISLABELED OR UNLABELED CONTAINERS

All companies must have a procedure for dealing with mislabeled or unlabeled containers. Ideally, such a situation will not occur; unfortunately labels fall off or are destroyed. The company procedure for addressing such a situation must be based on knowledge of the type chemicals used at the facility. At manufacturing or process facilities, which historically use only a limited variety of chemicals, initial identification of such containers can frequently be done from the type and color of the container. The initial identification can then be confirmed by an appropriate physical or chemical test.

Unlabeled containers in research facilities are significantly more difficult to address. The first rule is: **DO NOT MOVE THE CONTAINER!** The material may be shock sensitive, or the container may be in very poor condition. A thorough investigation and attempt to identify the material must be performed before any attempt to move it is allowed.

The second rule is: **DO NOT OPEN UNKNOWNS!** Some chemicals may be so friction sensitive that the energy of opening the container may cause ignition. The chemical may react with air. The chemical may cause severe health effects. Again, a thorough investigation and attempt to identify the material must be performed before any attempt to open it is allowed. If a reasonable estimate identity cannot be established, the container must be considered extremely unstable and reactive and must be handled accordingly.

INVENTORY CONTROL

Most chemical users prefer to leave the manufacturer's label on the product. Some companies attach an in-house label in addition to the original label. A recent trend is to place a bar code label on each container and use the bar code to track the container from the time it is received until the material is either completely used or disposed of.

IBM ARC

A 0076079

Figure 24 Bar code label used at the IBM Almaden Research Center for inventory control.

An example of such a bar code label is shown in Figure 24.[30] This particular application has become especially useful as regulatory agencies require increased reporting of chemical inventories.

TRAINING

Training employees to use chemical labels properly is not difficult. It involves going over the company policies and procedures that have been developed in accordance with appropriate Federal, state, and local regulations and with generally accepted work practices, and providing the employees with the in-house labels needed to implement the program. A particularly useful training device is a short exercise that was developed to provide employee involvement with the learning process. One version of the training exercise is presented in Table 11.

30 Software to support this inventory control bar code system is available from the IBM Corporation

TABLE 11
Label Information Exercise

MERCURY

UN 2809 **DANGER!**  **POISON!** CAS 743-99-76

TARGET ORGANS: SKIN, EYES, RESPIRATORY SYSTEM, KIDNEYS, CENTRAL NERVOUS SYSTEM

PRECAUTIONARY MEASURES: May be fatal or harmful if swallowed, absorbed through the skin, or inhaled. Keep in tightly closed container. Avoid breathing vapor, dust, or mist. Avoid contact with skin, eyes, or clothing. Use with adequate ventilation. Wash thoroughly after handling. When heated, emits very toxic fumes. In case of fire, use water spray, alcohol foam, dry chemical, or carbon dioxide extinguisher. Use appropriate personal protective equipment.

SYMPTOMS OF EXPOSURE: Coughing, dyspnea, bronchitis, tremor, insomnia, weakness, fatigue, headache, gastrointestinal distress, irritation of skin or eyes.

INSTRUCTIONS FOR IMMEDIATE FIRST AID IN CASE OF EXPOSURE: Call a Physician. If inhaled, remove to fresh air. If not breathing, give artificial respiration. If breathing is difficult, give oxygen. In case of contact, immediately flush affected area with water for at least 15 minutes. Medical intervention: treat symptoms.

SPILL RESPONSE PROCEDURES: Only appropriately trained personnel should respond to spills. Use spill response procedures. Maintain appropriate level of fire protection. Dispose of as required by federal, state, and local regulations.

CONTAINER HANDLING and STORAGE: Store in tightly closed container in a cool, dry location. Open containers carefully with adequate ventilation. Avoid storage or contact with incompatible materials.
REFER TO MATERIAL SAFETY DATA SHEET FOR MORE INFORMATION.

HG PRODUCTION, Inc.; Torrance, CA
(000) 111 - 2222

Using the sample label above, briefly answer the following questions:

1. What is the SIGNAL WORD?

   ____

2. List the major organs that may be affected by acute exposure to this chemical.

   ____

3. List the primary hazard(s) associated with this chemical.

   ____

   ____

4. The symbol to the right appears on this label. Briefly explain its meaning.

   ____

   ____

5. For the hazards listed in question 3, above, briefly describe the precautionary measure(s) recommended to avoid exposure to those hazards.

   ____

   ____

Another approach used to instruct employees in using the information from a label is to have them try to develop the precautionary information for a label. While this is a difficult task, as described throughout this book, we have developed an exercise that simply requires the participant to *strike-out* the unnecessary language. An example of this type training device is presented in Table 12 on the following pages.

TABLE 12
Chemical Label Preparation Exercise

You have been assigned the job of preparing a label for the new chemical **TETRAETHYL AWFULSTUFF**. Your Company Safety Office has prepared the attached Material Safety Data Sheet.

The attached **LABEL INFORMATION FORM** is designed for you to remove unnecessary information. To provide too much or too severe information is as much a safety problem as to provide too little information. Thus, for example, you must decide which **SIGNAL WORD** is most appropriate for this chemical.

TABLE 12 (continued)
LABEL INFORMATION EXERCISE

PRODUCT NAME:	
CHEMICAL NAME:	
SYNONYMS:	
CAS NUMBER:	
DOT IDENTIFICATION NUMBER:	

SIGNAL WORD: DANGER! WARNING! CAUTION! POISON!

TARGET ORGANS: skin, eyes, respiratory system, blood, CNS, bone marrow.

PRECAUTIONARY MEASURES: EXTREMELY FLAMMABLE. HARMFUL IF SWALLOWED, ABSORBED THROUGH SKIN, OR INHALED. CAUSES IRRITATION. Keep away from heat, sparks, flame. Keep in tightly closed container. Avoid breathing vapor. Avoid contact with eyes, skin, clothing. Use with adequate ventilation. Wash thoroughly after handling. In case of fire, use water spray, alcohol foam, dry chemical, or carbon dioxide extinguisher. Use appropriate personal protective equipment.

SYMPTOMS OF EXPOSURE: Irritation of the eyes, nose, respiratory system. Giddiness, dizziness, headache, nausea, staggered gait, or fatigue. Vomiting, diarrhea, narcosis, dermatitis, or gastrointestinal distress. Carcinogenic.

INSTRUCTIONS FOR IMMEDIATE AID IN CASE OF EXPOSURE: Call a physician. If swallowed, and if conscious, immediately induce vomiting. If inhaled, remove to fresh air. If not breathing, give artificial respiration. If breathing is difficult, give oxygen. In case of contact, immediately flush affected area with running water for at least 15 minutes. Medical intervention: treat symptoms of exposure.

SPILL RESPONSE PROCEDURES: Only appropriately trained personnel should respond to spills. Use proper response procedures and maintain appropriate level of fire protection. Dispose of as required by federal, state, and local regulations.

CONTAINER HANDLING and STORAGE: Use labeled flammable solvent container. Avoid contact with oxidizers and any other incompatible materials. Electrically ground container where necessary.

MSDS REFERENCE: Refer to Material Safety Data Sheet for more information about chemical.

MANUFACTURER:	
PHONE NUMBER:	
MSDS LOCATION:	

TABLE 12, continued

MATERIAL SAFETY DATA SHEET

I. PRODUCT IDENTIFICATION

TRADE NAME (AS LABELED):	TETRAETHYL AWFULSTUFF
CHEMICAL NAMES.:	1,3-CYCLOPENTADIENE
SYNONYMS:	CYCLOPENTADIENE
MANUFACTURER'S NAME:	ALL RIGHT CHEMICALS, Inc.
ADDRESS:	123 Main St. Yourtown, Yourstate
EMERGENCY PHONE:	000-000-0000
BUSINESS PHONE:	111-111-2222

II. HAZARDOUS INGREDIENTS

CHEMICAL NAME	CAS #	% v/v	EXPOSURE LIMITS IN AIR		
			TLV	PEL	OTHER
			ACGIH	OSHA	
1,3-CYCLOPENTADIENE	542-92-7	100	75 ppm	N/A	N/A

III. PHYSICAL PROPERTIES

VAPOR DENSITY: 1.34

SPECIFIC GRAVITY: 0.8

SOLUBILITY IN WATER: slight

EVAPORATION RATE (BuAc 1): 0.6

MELTING POINT or RANGE: 20°F

BOILING POINT: 107°F

VAPOR PRESSURE, mm Hg @ 20 C: 7 mm

APPEARANCE and COLOR: Pale yellow liquid.

HOW TO DETECT THIS SUBSTANCE (warning properties): "Organic solvent" odor.

IV. FIRE and EXPLOSION

FLASH POINT,°F (method): 85°F cc

AUTOIGNITION TEMPERATURE, °F: 350°F

FLAMMABLE LIMITS in air by volume: lower 1.2% upper 7.1%

FIRE EXTINGUISHING MATERIALS:

water spray: for cooling	carbon dioxide: yes	foam: yes
dry chemical: yes	halon: yes	other:

SPECIAL FIRE FIGHTING PROCEDURES: On small fires, use dry chemical or carbon dioxide. On large fires, use standard fire fighting agents. Cool containers with water from maximum distance. Continue cooling after fires have been extinguished. Do not approach ends of horizontal tanks. Move exposed containers from fire area, if without risk. Withdraw immediately upon hearing rising sounds from vent device or discoloration of tank due to fire.

UNUSUAL FIRE and EXPLOSION HAZARDS: Heated containers may rupture violently and produce flying missiles.

V. HEALTH HAZARD INFORMATION

SYMPTOMS OF OVEREXPOSURE BY ROUTE OF EXPOSURE:

INHALATION: Poisonous if breathed. Breathing will become labored very quickly. Extremely irritating to nose and throat.

CONTACT WITH SKIN or EYES: May cause severe burns to skin and eyes. Irritation at clothing pressure points; eyes will feel dry and will burn.

SKIN ABSORPTION: May be toxic through skin absorption. Skin of face, hands, neck, and or the exposed surfaces will redden and burn.

INGESTION: Liquid may cause death if consumed. Nausea and vomiting may follow ingestion.

INJECTION: Toxic, may cause death in large amounts.

HEALTH EFFECTS OR RISKS FROM EXPOSURE (An explanation in lay terms):

ACUTE: May cause breathing difficulty, dizziness, and headaches.
CHRONIC: Prolonged headaches, disorientation, breathing difficulties.

FIRST AID EMERGENCY PROCEDURES:

If spilled on skin or eyes, immediately begin decontamination with running water. Remove exposed or contaminated clothing, taking care not to contaminate eyes. Victim and rescuers must seek immediate medical attention. If chemical is in eyes, open victim's eyes while under gentle running water. Use sufficient force to open eye lids. Have victim "roll" eyes. Minimum flushing is for 15 minutes. If chemical is inhaled, remove victim to fresh air and use artificial respiration to support vital functions. Remove or cover gross contamination to avoid exposure to rescuers. **If chemical is swallowed do not induce vomiting.**

Victim of chemical exposure and all rescuers must be taken for medical attention. Take copy of label and MSDS to physician or health professional with victim.

SUSPECTED CANCER AGENT?

This product's ingredients ARE NOT found on the following lists:
FEDERAL OSHA Z LIST National Toxicology Program International Agency for Research on Cancer CAL/OSHA

MEDICAL CONDITIONS AGGRAVATED BY EXPOSURE: Chronic lung disease, psoriasis.

RECOMMENDATIONS TO PHYSICIANS: Treat symptoms, monitor liver function.

VI. REACTIVITY DATA

STABILITY: Stable.
CONDITIONS TO AVOID: Heat, sparks, flame.
MATERIALS WITH WHICH SUBSTANCE IS INCOMPATIBLE: Oxidizers, strong acids, strong caustics, aluminum based alloys.
HAZARDOUS POLYMERIZATION: May occur.
CONDITIONS TO AVOID: Avoid extreme heat and contact with incompatible materials.

VII. SPILL, LEAK, and DISPOSAL INFORMATION

SPILL and LEAK RESPONSE: Uncontrolled releases should be responded to using training people and preplanned procedures. Proper protective equipment (PPE) should be used. In case of a spill, clear the affected area, protect people, and respond with trained personnel. Minimum PPE should be hard hat with face shield, splash goggles, respiratory protection (Self Contained Breathing Apparatus, unless conditions warrant lower protection), hand and body protection, and chemical protective boots.

PREPARING WASTES FOR DISPOSAL: Waste disposal must be in accordance with appropriate federal, state, and local regulations. This chemical, if unaltered by the handling, may be disposed of by recycling or incineration.

VIII. SPECIAL HANDLING INFORMATION

VENTILATION and ENGINEERING CONTROLS: Use adequate ventilation. Do not use in enclosed spaces. Do not use in poorly ventilated areas.
RESPIRATORY PROTECTION: None normally required unless ventilation inadequate. Consult with qualified industrial hygienist to determine actual respiratory protection needed.
EYE PROTECTION: Chemical splash glasses or goggles.
HAND PROTECTION: Sol-vex (nitrile) gloves recommended if skin contact possible.
BODY PROTECTION: Normal work clothes appropriate for task. Coated PVC or Saranex aprons suggested.
WORK PRACTICES and HYGIENE PRACTICES: Follow the rules of chemical safety. Wash hands after use.
STORAGE and HANDLING PRACTICES: Store in flammable liquid area. Keep containers closed, except when in use. Store in well ventilated, cool, dry location. Appropriate fire protection must be available.
PROTECTIVE PRACTICES DURING MAINTENANCE OF CONTAMINATED EQUIPMENT: Level C protection is suggested. Provide task safety review prior to commencing work.

IX LABELING

LABELING (Precautionary Statements): THIS IS YOUR ASSIGNMENT!
DOT LABEL(S): FLAMMABLE LIQUID; UN1993

PREPARED BY
CHEMICAL SAFETY ASSOCIATES, Inc. 9163 Chesapeake Dr., San Diego, CA 92111
619/565-0302 DATE OF PREPARATION: 1/1/86

The employer is responsible to provide the employee with the tools needed for a safe work place. It is up to the employee to work safely. Adequate training will help the employee develop the habits needed to handle chemicals safely.

MATERIAL SAFETY DATA SHEETS

Chapter 9

PRECAUTIONARY LABELS FOR CHEMICAL CONTAINERS

INTRODUCTION

This book concerns the primary source of necessary information about chemicals - the precautionary label. The system OSHA designed to convey precautionary information to workers depends upon the use of both the label and the MSDS. Thus, it is appropriate to include information on the MSDS in this book.

The Chemical Manufacturer's Association (CMA), in conjunction with the American National Standards Institute (ANSI) and with the concurrence of both OSHA and the EPA has prepared a guidance for the Material Safety Data Sheet. The remainder of this chapter explains the information that typically appears on an MSDS; using an example MSDS for a totally fictitious product - **PR-20**.

The Material Safety Data Sheet is the primary source of technical information concerning safe handling procedures and health affects of a chemical. According to the OSHA Hazard Communication Standard (HCS) regulations, a Material Safety Data Sheet must be readily available near the work station for every hazardous chemical used. Employees must know how to obtain a Material Safety Data Sheet and must be trained in how to interpret them. Because of the central role a Material Safety Data Sheet plays (along with labels), in providing the **key to chemical safety**, it is important that you understand how to read them. In this section, we will refer to the Material Safety Data Sheet for the compound "**PR-20,**" which is located at the end of this section. Each part of this sample Material Safety Data Sheet will be explained in nontechnical terms, designed for the average chemical user.

The Chemical Manufacturer's Association and the American National Standards Institute (ANSI), with the support of both OSHA and the EPA, have developed a Material Safety Data Sheet format, but the format *is not required*. Any form or format can be used as long as the minimum required information is present. Because the CMA/ANSI format will become the standard of the industry, the new format has been used for this section. This example has more information than the minimum required.

PART I *What is the material and what do I need to know in an emergency?*

1. PRODUCT IDENTIFICATION

The compound is known by the name "PR-20." It may have several other names associated with it, but this is the only name generally used in the work place. If other names were common, these would also be listed in this section. This section also provides the manufacturer's name, address, and telephone number. If a unique chemical species description were applicable, the chemical name and formula would be included in this section.

2. COMPOSITION AND INFORMATION ON INGREDIENTS

This section lists the major components and their associated hazards. Any component in less than 1% concentration need not be listed, unless it is a carcinogen (causes or is suspected to cause cancer), in which case it must be listed if present in a concentration of 0.1% or greater. This section also includes the AMERICAN CONFERENCE OF GOVERNMENTAL INDUSTRIAL HYGIENISTS (ACGIH) **THRESHOLD LIMIT VALUE (TLV)** and **SHORT-TERM EXPOSURE LIMIT (STEL)**, which are guides to how toxic the chemical may be. The equivalent OSHA values, the **PERMISSIBLE EXPOSURE LIMIT (PEL)**, the **STEL**,and the **IMMEDIATELY DANGEROUS TO LIFE AND HEALTH (IDLH)** values are also included, if such values have been published. The TLV and PEL are guidelines that tell you how much exposure to a chemical one can tolerate for an 8 hour work day, 5 days per week, for a normal working career and not suffer ill-effects. In the work place, we try to use ventilation, equipment design, and other methods to maintain the concentration of fumes and vapors well below the TLV. Since the TLV measures a concentration, it must be expressed in the terms of some unit of concentration. The most common units are the number of parts of the compound in 1,000,000 parts of air (ppm) or the weight in milligrams (mg) in a cubic meter (m^3) of air. In either case, the lower the number, the more toxic the compound. In order to assist in interpreting this Material Safety Data Sheet, the table below presents some values for common work place chemicals.

TLVs OF SOME COMMON CHEMICALS		
CHEMICAL	TLV (1991)	
	ppm	mg/m^3
GASOLINE	300	900
ACETIC ACID	10	25
CARBON DIOXIDE	5000	9000
STODDARD SOLVENT	100	525

A complete listing of TLVs is updated and published yearly by the American Conference of Governmental Industrial Hygienists.

Referring to the PR-20 MSDS, we see **methylene diphenylisocyanate** has a TLV marked with a "C" to indicate that it is a ceiling value. A ceiling value is a concentration level to which no worker should be exposed, under any circumstances, for any period of time. Another component of PR-20, **cyclohexanone** has the notation "skin" which indicates that this compound may absorb through the skin and thus increase the true body exposure above that of exposure from the respiratory route of entry.

Another value sometimes reported is the **PEL,** which is of identical meaning to the TLV, but is published by OSHA and is a value that can be enforced by regulation.

Because no TLV has been established for **PR-20**, a prudent company will use engineering controls and/or work place procedures to maintain the concentration of each component below the published TLV.

Other information frequently found in this section is the "Chemical Abstract Service Number" (CAS #), which provides a key to computer based literature and technical information concerning the chemical. The composition, expressed in an appropriate unit, may also be included. OSHA does not require this information, but many manufacturers choose to include it, for completeness.

3. HAZARD IDENTIFICATION

This section of the MSDS is designed to provide information needed to recognize if an *overexposure* to the chemical has occurred. Symptoms of overexposure, by route of entry, for both acute and chronic exposures, are described in sufficient detail to allow a person to recognize symptoms. This section also contains a succint Emergency Overview to provide emergency personnel with clear, immediate information.

PART II *What should I do if a hazardous situation occurs?*

This entire section of the MSDS is dedicated to telling you how to handle an emergency. The section addresses first-aid, fires and explosions, and chemical spills or releases.

4. FIRST-AID MEASURES

This section will provide you with specific information needed in the event a person gets the chemical *in them or on them.* If specific antidotes or medical treatments exist, this section will present that information. This section will also provide information concerning decontamination of the skin or eyes.

5. FIRE-FIGHTING MEASURES

The section on fire-fighting measures will provide information on the flammability of the chemical as well as specific information regarding the proper fire extinguishing medium for use on a fire involving the chemical.

This section involves the use of some very technical language which you will need to understand to appreciate the physical/chemical properties important to fire suppression and prevention.

FLASH POINT: A flash point is the lowest temperature at which a liquid gives off a vapor in sufficient concentration to ignite when given a source of ignition. Self-sustained combustion of the material may or may not continue after ignition of the vapors. Some examples of flash points include:

Carbon disulfide	-30 °C
Acetone	-18 °C
Methyl alcohol	11 °C
Nitromethane	35 °C
Benzoyl alcohol	100 °C

IGNITION TEMPERATURE: Ignition temperature is the minimum temperature required to initiate self-sustained combustion of a material. This temperature may be the same as the flash point but is usually slightly higher.

AUTOIGNITION TEMPERATURE: The autoignition temperature is the lowest temperature at which a material will spontaneously ignite and burn. In this situation, heat is the only source of ignition. For most materials, the autoignition temperature tends to be high (several hundred degrees C). However, some material, such as white phosphorus, will autoignite at temperatures close to room temperature.

FLAMMABILITY LIMITS: Flammability limits are the lowest and highest concentration of vapors in air that will produce a flash fire when an ignition source is provided. Its unit of measurement is percent by volume in air. At a concentration too low to ignite, the mixture is too "lean" to burn. At a concentration too high to burn, the mixture is too "rich" to burn. Some examples of flammability limits include:

Acetic acid, conc.	5.4 - 16.0 %
Acetylene	2.5 - 81.0 %
Ammonia	16.0 - 25.0 %
Gasoline	1.4 - 7.6 %
Hydrogen	4.0 - 75.0 %

Now, let's look again at the information given in the MSDS for PR-20. This material has a flash point of 80°F with a flammability range of 1.3 to 7.1%. Since the flash point is the temperature below which vapors from the liquid will not ignite when provided with a source of ignition, this says that fire hazard can be reduced by keeping the room cool. Gasoline, for reference, has a flash point of about - 45°F. If PR-20 were the fuel for your car, could you start it most mornings?

LOWER FLAMMABILITY LIMIT (LFL): This is the concentration below which the vapors are *too lean* to support combustion.

UPPER FLAMMABILITY LIMIT (UFL): This represents the concentration above which the fumes are *too rich* to support combustion. PR-20 has similar LFL and UFL properties to gasoline.

By combining this information with vapor density information you should now be able to decide if it is "safe" to work near an open 25 gallon container of PR-20 with a common electric motor on the floor.

Including the National Fire Protection Association (NFPA) "704" system information on the chemical or chemical mixture on the MSDS is common practice. The information contained in the diamond pertains to three major hazards the chemical may present if involved in a fire: health effects, flammability, and reactivity with water or air. In addition, special items, such as radioactivity, biohazard, or oxidizer is included in the "other" section. The potential hazard in each field is rated on a scale of zero (minimal hazard) to 4 (extreme hazard). This product is rated 3-0-1, which indicates a high overall degree of hazard. While this warning system is aimed at firefighters, the information is useful to anyone who handles this product as another clear indication of potential hazards. If this product is stored in bulk, the storage tank should be posted with the NFPA diamond, in such a manner that it is clearly visible from any direction of approach at a distance of 50 meters.

The remainder of this section in the MSDS tells our safety engineers about the type of fire extinguisher to provide in locations where this chemical is being used and tells fire fighters about toxic fumes given off by fires involving PR-20. This information implies that even a beginning (incipient) fire involving PR-20 could release a very toxic gas - hydrogen cyanide from methylene diphenylisocyanate (MDI) and that Self-Contained Breathing Apparatus (SCBA) must be used.

6. ACCIDENTAL RELEASE MEASURES

As with many Material Safety Data Sheets, this section is very general. It tells you some general things to do in case of a spill. Ideally, your company will have a spill preplan in place for every chemical used.

PART III *How can I prevent hazardous situations from occurring?*

7. HANDLING AND STORAGE

The information in this section has been significantly expanded over that which was commonly found on older MSDSs. This section now provides very specific instructions for how to store the chemical properly and details for handling the material. In addition, any specialized information needed with respect to maintenance activities must be addressed in this section, to prevent accidents of the type that have occurred in many chemical plants during shut-downs.

It is important that this section of an MSDS be reviewed prior to initiation of any maintenance activities. This will minimize the risk of accidents or injuries.

8. EXPOSURE CONTROLS - PERSONAL PROTECTION

This section has been expanded and is designed to inform employees of the engineering controls that can be used to minimize exposures. The section must be divided into subsections that address engineering controls, administrative practices, and personal protective equipment. The latter must address each part of the body that requires protection.

9. PHYSICAL AND CHEMICAL PROPERTIES

This section contains a wealth of important information, if you can understand it. Before we begin, some definitions are in order.

MELTING POINT: The melting point of a material is the temperature at which a solid substance changes to a liquid state. If a material has a melting point below the ambient temperature, it will be a liquid. If it has a melting point above the ambient temperature, you will find it in the solid phase. Pure chemicals will melt at a specific temperature while chemical mixtures tend to melt over a temperature range of several degrees. Some materials will not melt at all but will simply decompose if the temperature gets high enough. Some examples of melting points include:

MELTING POINTS OF COMMON MATERIALS

Sodium chloride	804 °C
Moth balls (p-chloro benzene)	53.5 °C
Phenol	43 °C
Ice (water)	0 °C
Sodium hydroxide	318 °C

BOILING POINT: The boiling point of a material is the temperature at which the vapor pressure of a liquid equals the surrounding pressure. The boiling point is expressed in degrees (temperature), at sea level pressure (760 mm Hg). Some examples of boiling points include:

BOILING POINTS OF COMMON MATERIALS

Propane	-42 °C
Anhydrous ammonia	-32 °C
Butane	- 1 °C
Ether	35 °C
Allyl chloride	45 °C
Ethylene glycol	387 °C

Now let us look specifically at the information for PR-20. This mixture boils 13°C below the boiling point of water. Thus, if it is being heated, it will come to a boil much more quickly than water. This is also reflected by its vapor pressure. At 20°C, the vapor pressure of water is 18 mm Hg, while PR-20 has a vapor pressure twice of that of water. The general rule is: the lower the boiling point, the higher the vapor pressure.

VAPOR DENSITY: The vapor density is the ratio of the weight of a vapor or gas compared to the weight of an equal volume of air. Materials lighter than air have vapor densities less than 1.0, such as acetylene, methane, and hydrogen. Materials heavier than air have vapor densities greater than 1.0, such as propane, hydrogen sulfide, ethane, butane, chlorine, and sulfur dioxide.

PR-20 vapors are 3.4 times as heavy as air. Thus, if one is performing a task that involves allowing vapors from this material to escape, they will sink and hug the ground, unless very good ventilation is used.

VAPOR PRESSURE: This is the pressure exerted by a saturated vapor above its own liquid in a closed container at some specified temperature. The vapor pressure of a material increases as temperature increases. The units of measurement are given in mm Hg (millimeters of mercury). A liquid boils when its vapor pressure is equal to the ambient air pressure.

EVAPORATION RATE: The rate at which a material will vaporize (evaporate) when compared to the known rate of vaporization of a standard material. The designated standard material is usually normal butyl acetate (NBUAC or n-BuAc), with a vaporization rate designated as 1.0. Evaporation rates of other material can then be classified as follows:

FAST: Evaporation rate > 3.0
- Methyl ethyl ketone = 3.8
- Acetone = 5.6
- Hexane = 8.3

MEDIUM: Evaporation rate > 0.8 and < 3.0
- 95% Ethyl alcohol = 1.4
- Naphtha = 1.4
- Methyl isobutyl ketone = 1.6

SLOW: Evaporation rate < 0.8
- Xylene = 0.6

Isobutyl alcohol = 0.6

Water = 0.3

Mineral spirits = 0.1

If we completely evaporate some PR-20, a residue will remain, which could amount to as much as 20 - 40% of the volume initially allowed to evaporate. The evaporation process would occur some 27 times more rapidly than the same volume of water, since PR-20 evaporates about 4.5 times as fast as normal butyl acetate (nBuAc), which evaporates some 6 times faster than water.

SPECIFIC GRAVITY: Specific gravity is the ratio of the weight of a material compared to the weight of an equal volume of water. Since specific gravity is a ratio it is a unitless number. Insoluble materials with a specific gravity of less than 1.0 will float on water. Insoluble materials with a specific gravity greater than 1.0 will sink in water. Most, but not all, flammable liquids have a specific gravity less than 1.0. Some examples of specific gravities are as shown:

SPECIFIC GRAVITIES OF SOME COMMON MATERIALS

Acetone	0.8
Toluene	0.9
Hydrochloric acid	1.2
Naphthalene	1.1
Potassium chlorate	2.3

SOLUBILITY: Solubility is the property of being soluble in some solvent, usually water. Solubility is based on an observation on how much material, by weight, dissolves in a stated quantity (weight or volume) of liquid. This value is expressed as the mass in grams of a material that will dissolve in 100 grams of solvent, at room temperature.

Notice that PR-20 is insoluble in water and has a specific gravity that is 1.37 times that of water. Thus, if PR-20 is spilled into water it would sink, and our spill cleanup would have to include the bottom of the water system. Obviously, if this were a stream, it would be very costly to cleanup and decontaminate.

Putting all of this together, the physical data indicates to us that PR-20 will evaporate very rapidly, leaving a residue. The fumes from evaporation will sink, and, if the chemical gets into water, the liquid will sink.

10. STABILITY and REACTIVITY

The stability and reactivity data section tells us what to expect from this material during storage or if mixed with other chemicals. Thus, PR-20 is a stable compound at room temperature that will not self-react (polymerize) to generate other products and potentially dangerous heat or by-products. However, polymerization may occur at elevated temperatures. Further, if exposed to almost any contaminant - alcohols, strong bases, or strong oxidizers, the isocyanate component may react and potentially cause problems. This is what may have occurred during several well publicized chemical accidents involving isocyanates.

This section reminds us that the mixture is *flammable* and to follow proper safety rules for such flammable materials.

PART IV *Is there any other useful information about this material?*

This is a new part for the MSDS. The information contained in the sections making up this part of the MSDS is aimed at providing additional information, such as regulatory information and more health information, which is needed for proper chemical management and for environmental protection. The sections contained in this part were added to the MSDS as a result of discussions among the agencies, the Chemical Manufacturer's Association, and other interested parties.

11. TOXICOLOGICAL INFORMATION

This section introduces a variety of information to help you understand effects of exposure to PR-20. **All chemicals can be handled safely, and your body can readily adapt to low level exposures to many toxic compounds.** This section of the Material Safety Data Sheet will provide you with an interpretation of "low level" for this compound.

This Material Safety Data Sheet specifically tells whether or not any reports suggesting that this material, or a component of this material, may cause cancer. The presence of a published scientific report does not prove that a chemical is a carcinogen but should alert you to be particularly careful when using the product. The lists referred to are:

FEDERAL OSHA Z LIST: This is the list of Toxic and Hazardous Substances published by OSHA. It is found in Title 29 of the Code of Federal Regulations, in Subpart Z.

NTP: This is the National Toxicology Program, which is a federal government program designed to evaluate the toxicity of various chemicals.

IARC: The International Agency for Research on Cancer conducts and reports results from research on cancer. It is internationally recognized as a major compilation of data related to chemical carcinogens.

CAL/OSHA LIST: CAL/OSHA publishes extensive data, taken from many sources about potential carcinogens.

In addition to indicating what agency or agencies list a component of the product as a potential carcinogen, the MSDS should also indicate the basis for the listing and, if available, the rating assigned to the chemical. This information is clearly indicated on the MSDS for PR-20.

This section also includes some information that will allow you to determine if other medical conditions could be aggravated by exposure to this material. This section should be very specific, if possible, citing specific diseases or conditions that could be made worse by acute or chronic exposure.

The final information in this section may be quite technical since it is directed specifically at physicians.

12. ECOLOGICAL INFORMATION

This section of the MSDS is to provide the chemical user with information on the impact of the chemical on the ecosphere, if it is released into the environment. Not much data are available on the environmental stability of chemicals or on the impact of chemicals on fish, plants, or animals. Initially, it is to be expected that this section will be very generic; eventually, more specific information will become available and will be incorporated into MSDSs.

13. DISPOSAL CONSIDERATIONS

Disposal information has traditionally been placed on MSDSs. The new MSDS format places the disposal information in a section by itself. It is not likely that the information provided will be more specific than is given in the PR-20 example. Preparers of MSDSs will still leave the actual waste identification and listing as a generator responsibility.

The EPA Waste Numbers provided reflect the material as manufactured, if it becomes a waste without any modification. If processing changes its chemical or physical properties, the EPA Waste Numbers could change.

14. TRANSPORTATION INFORMATION

With the complete revision of the Hazardous Materials Shipping Regulations by the U.S. DOT (HM-181), it is important that the MSDS provide accurate information regarding shipping. Most, but not all, chemicals for which an MSDS is required will ship as DOT Hazardous Materials. In order to understand the information provided, you must again understand the language used. Thus, we have provided some definitions.

Proper Shipping Name is the name and description assigned to a chemical by the Department of Transportation in 49 CFR 172.101 - the Hazardous Materials

Shipping Table. If this information is not correctly determined, none of the following information can be correct. For the example MSDS, the Proper Shipping Name is determined to be *Flammable liquid, poisonous, n.o.s. (Cyclohexanone, Methylene diphenyl isocyanate).* The technical names given in the parenthesis is required since the Proper Shipping Name is a generic name, with the "n.o.s.," (that is, "not otherwise specified") notation included.

Hazard Class Number and Description is the number, from 1 to 9, that identifies the specific hazard class to which the chemical belongs. In our example, PR-20 belongs to Hazard Class 3, which is described as "Flammable Liquid."

DOT Identification Number, which is frequently referred to as the **UN Number** is a six character alpha-numeric code which begins with the letters UN (United Nations) or NA (North American) and is followed by four numbers. The DOT Identification Number is placed on shipping papers and, when appropriate, on the side of trucks. It serves to identify clearly the commodity in transit in the event of an emergency and informs emergency responders on proper initial response procedures. For the example compound, the DOT Identification Number is UN 1992.

Packaging Group is the new DOT Performance Oriented Packaging, which replaces the specification packaging of the old DOT. The Packaging Group represents one of three levels of packaging integrity. The greater the hazard associated with the material, the greater the integrity of the packaging required by DOT. Packaging Group I (PG I) is the strongest; PG III is the least strong. The first is the correct choice for the example compound.

Emergency Response Guide Number is a reference to the DOT emergency guide number, which corresponds to the Proper Shipping Name selected for the shipment. The Emergency Response Guide Number will instruct emergency responders as to the initial procedures to follow in the event the shipment is involved in an accident. Guide Number 28 is the correct choice in this particular case.

15. REGULATORY INFORMATION

Section 15 is a new section on the MSDS. It contains a compilation of various types of regulatory information, primarily environmental. Chemicals are regulated by many state, as well as federal, regulations.

Chemicals that are not on the Toxic Substance Control Act (TSCA) Inventory can only be used in the United States under very specific circumstances. This section will alert the chemical user if there is a potential TSCA conflict.

The list of state regulations addresses a wide variety of specific state regulations of which the chemical user should be aware. This section will alert the chemical user to examine specific state regulations for more information.

The California Safe Drinking Water Act, otherwise known as Proposition 65, requires specific precautionary wording on certain chemical labels and in locations where certain chemicals are used. This section will alert the chemical user located in California to a need to comply with Proposition 65 when using this material.

Under the Comprehensive Environmental Response, Compensation, and Liability

Act (CERCLA or Superfund), a person must report to the National Response Center in the event of the offsite release or threat of offsite release of a reportable quantity of the chemical.

It is now common for the precautionary information from the chemical label to be placed on the MSDS. This assists medical care providers in the event the chemical is ingested or spilled on someone. It also brings all necessary precautionary information together on the MSDS.

16. OTHER INFORMATION

OSHA requires that the name and address of the preparer of the MSDS and the date of preparation appear on the MSDS.

MATERIAL SAFETY DATA SHEET

PART I *What is the material and what do I need to know in an emergency?*

1. PRODUCT IDENTIFICATION

TRADE NAME (AS LABELED): **PR-20**

CHEMICAL NAMES, COMMON NAMES: NONE/MIXTURE
SYNONYMS: NONE/MIXTURE

MANUFACTURER'S NAME: **AAA CHEMICAL CORPORATION**
ADDRESS: **555 MAIN STREET, YOURTOWN, YOURSTATE**
EMERGENCY PHONE: **000/111-1111**
BUSINESS PHONE: **000-222-2222**

2. COMPOSITION and INFORMATION ON INGREDIENTS

CHEMICAL NAME	CAS #	% w/w	EXPOSURE LIMITS IN AIR					
			ACGIH		OSHA			OTHER
			TLV mg/m³	STEL mg/m³	PEL mg/m³	STEL mg/m³	IDLH mg/m³	
LEAD-SILICOCHROMATE PIGMENT*	N/A	12	.15	-	0.15	-	-	-
METHYLENE DIPHENYL ISOCYANATE*	101-68-8	7	0.2C 0.02 ppm	-	0.2C 0.02 ppm	-	-	-
CYCLOHEXANONE	108-94-1	25	25 ppm (skin)	-	25p pm	-	-	-
TRICHLORO-ETHYLENE*	79-01-6	40	50	-	50	-	-	-
CHLOROBENZENE*	108-90-7	7	75	-	75	-	-	-
REMAINDER IS INERT AND NON-HAZARDOUS	-	9	-	-	-	-	-	-

N/A = not applicable - = no information available

3. HAZARD IDENTIFICATION

<u>EMERGENCY OVERVIEW</u>: This brown, cloudy liquid has an intense solvent odor. It is flammable and may release toxic gases or metal fumes during a fire. Emergency personnel should exercise caution in the event of an emergency involving large amounts of this product.

<u>SYMPTOMS OF OVEREXPOSURE BY ROUTE OF EXPOSURE</u>

<u>INHALATION</u>: Isocyanates are lachrymators. Exposure to mists or vapors may cause eyes and mucous membrane irritation, headache, drowsiness, or narcosis. Extreme exposure can cause central nervous system (CNS) disturbance, respiratory or cardiac failure, and death.

<u>CONTACT WITH SKIN or EYES</u>: Skin contact can cause dermatitis or burns. Eye contact can cause injury, burning, or blindness.

<u>SKIN ABSORPTION</u>: Product may skin absorb, particularly through breaks in skin. Symptoms will be similar to those of inhalation exposure.

<u>INGESTION</u>: Nausea, diarrhea, and symptoms similar to those of inhalation.

<u>INJECTION</u>: Burning at point of injection; symptoms similar to those of inhalation may develop.

<u>HEALTH EFFECTS OR RISKS FROM EXPOSURE. An explanation in lay terms</u>.

ACUTE:

Overexposure will cause victim to feel ill, drowsy, or disoriented. Breathing may be difficult, heart may beat rapidly, victim may lose consciousness.

CHRONIC:

Skin irritation may develop after prolonged contact. Liver problems may develop after prolonged exposure above the TLV.

PART II

What should I do if a hazardous situation occurs?

4. FIRST-AID MEASURES

If spilled on skin or eyes, <u>immediately</u> begin decontamination with running water. Remove exposed or contaminated clothing, taking care not to contaminate eyes. Victim and rescuers must seek immediate medical attention. If the product is in eyes, open victim's eyes while under gentle running water. Use sufficient force to open eye lids. Have victim "roll" eyes. <u>Minimum</u> flushing is for 15 minutes. If chemical is inhaled, remove victim to fresh air and use artificial respiration to support vital functions. Remove or cover gross contamination to avoid exposure to rescuers.

If chemical is swallowed, CALL PHYSICIAN OR POISON CONTROL CENTER FOR MOST CURRENT INFORMATION. If professional advice is not available, DO NOT induce vomiting. Never induce vomiting or give diluents (milk or water) to someone who is unconscious, having convulsions, or who cannot swallow.

Victim of chemical exposure and all rescuers must be taken for medical attention. Take copy of label and MSDS to physician or health professional with victim.

5. FIRE-FIGHTING MEASURES

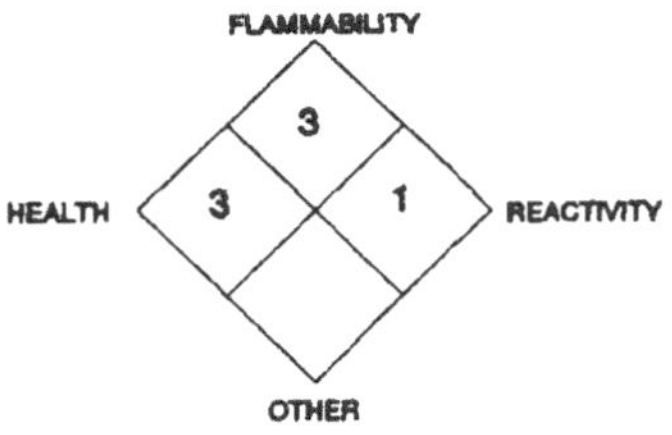

FLASH POINT, °F (method): 80°F(CC)
AUTOIGNITION TEMPERATURE, °F: 330°F
FLAMMABLE LIMITS in air by volume, %:
lower 1.3 upper 7.1

FIRE EXTINGUISHING MATERIALS:

Water Spray: NO Carbon Dioxide: OK Foam: OK
Dry Chemical: OK Halon: OK

SPECIAL FIRE FIGHTING PROCEDURES: Exposure to fire can create toxic fumes and vapors including methylene diphenyl isocyanates and nitrogen oxides. Incipient fire responders should wear eye protection. Structural fire fighters must wear self-contained breathing apparatus and full protective equipment.

UNUSUAL FIRE and EXPLOSION HAZARDS: Flammable liquid fire hazards. Toxic combustion products may be released. When heated to decomposition, this product will emit toxic fumes containing oxides of nitrogen, lead, and chrome.

6. ACCIDENTAL RELEASE MEASURES

SPILL and LEAK RESPONSE: Uncontrolled releases should be responded to by trained personnel using preplanned procedures. Proper protective equipment should be used. In case of a spill, clear the affected area, protect people, and respond with trained personnel. Minimum personal protective equipment should be rubber gloves and rubber apron, splash goggles, or safety glasses. Respiratory protection should be used. Absorb spilled liquid with polypads or other appropriate materials. Return usable liquid to process. Decontaminate area thoroughly. Place all spill residue in a double plastic bag, in appropriate sealed drum. Dispose of in accordance with federal, state, and local hazardous waste disposal regulations.

PART III *How can I prevent hazardous situations from occurring?*

7. HANDLING and STORAGE

WORK PRACTICES and HYGIENE PRACTICES: Avoid getting chemicals ON YOU or IN YOU. Wash hands after handling chemicals. Do not eat or drink while handling chemicals. Follow SPECIFIC USE INSTRUCTIONS supplied with product. This product contains lead. Comply with the OSHA Lead Standard.

STORAGE and HANDLING PRACTICES: Store in closed containers in well ventilated area. Do not expose to sources of ignition. Ground drums. Maintain storage temperature between 60° and 90°F. Segregate by secondary containment from other chemicals. Do not store near acids, caustics, oxidizers, or alcohols.

PROTECTIVE PRACTICES DURING MAINTENANCE OF CONTAMINATED EQUIPMENT: Follow practices indicated above. Make certain application equipment is locked and tagged-out safely. Always use this product in areas where adequate ventilation is provided. Decontaminate equipment by thoroughly rinsing to remove all traces of lead or chrome. Collect all rinsates and dispose of according to applicable local, state, or federal procedures.

8. EXPOSURE CONTROLS - PERSONAL PROTECTION

VENTILATION and ENGINEERING CONTROLS: Use with adequate ventilation. Use a mechanical fan or vent area to outside. Area should have at least three air changes per hour.

RESPIRATORY PROTECTION: Industrial hygiene monitoring should be performed to verify actual exposure levels. Comply with the requirements of the OSHA lead standard for high or unknown air borne concentrations of lead or chrome. Use an air purifying respirator with organic vapor cartridge and high efficiency particulate filter only when actual concentration of airborne contaminant has been determined by monitoring.

EYE PROTECTION: Splash goggles or safety glasses. A full face shield should be used when handling large volumes of the material (i.e. dispensing from drums).

HAND PROTECTION: Use appropriate gloves. Neoprene, and nitrile (Solvex™) gloves are generally acceptable. Check gloves for leaks. Wash hands after removing gloves.

BODY PROTECTION: Use body protection appropriate for task. Chemical protective clothing made from Tyvek™ or similar impermeable material are generally acceptable.

9. PHYSICAL and CHEMICAL PROPERTIES

VAPOR DENSITY:	3.4 (air = 1)	EVAPORATION RATE (NBUAC=1):	4.5
SPECIFIC GRAVITY:	1.37	MELTING POINT or RANGE: °C	N/A
SOLUBILITY IN WATER:	slight	BOILING POINT: °C	87
VAPOR PRESSURE, mm Hg @ 20 °C:	36		

APPEARANCE and COLOR: Brown, cloudy liquid, chlorinated solvent odor.

HOW TO DETECT THIS SUBSTANCE (warning properties): sweet smell of "dry cleaning solvent" (chlorinated solvent odor).

10. STABILITY and REACTIVITY

STABILITY: Unstable at elevated temperatures.

CONDITIONS TO AVOID: Metal contamination, heat, sparks, flame, oxidizers, acids, caustics.

MATERIALS WITH WHICH SUBSTANCE IS INCOMPATIBLE: Product is not compatible with acids, caustics, oxidizers, alcohols.

HAZARDOUS POLYMERIZATION: May occur if contaminated or heated.

PART IV *Is there any other useful information about this material?*

11. TOXICOLOGICAL INFORMATION

SUSPECTED CANCER AGENT
This product's ingredients are found on the following lists:

FEDERAL OSHA Z LIST	NTP	IARC	CAL/OSHA
YES	YES	YES	YES

Basis for Listing:

Federal OSHA regulates both lead and chrome compounds as potential human carcinogens. Refer to 29 CFR 1910.1025 and 1000 for OSHA required handling procedures.

The National Toxicology Program considers lead compounds as Group 2B. It considers chromates as Group 1.

The International Agency for Research on Cancer considers lead compounds as Group 2B. It considers chromates as Group 1.

CAL/OSHA has listed both lead and chrome compounds as human carcinogens.

Dermal Exposure: This product can cause an acute reaction as a result of dermal exposure. Chronic illness can result as a result of dermal exposure.

Ingestion Exposure: This product will cause a severe acute response upon ingestion. Residual lead or chromium may remain in the body, contributing to a potential chronic exposure to these metals.

Inhalation Exposure: The vapor pressure of this material is sufficiently high that acute exposure to the volatile ingredients may occur. The exposure may be both acute or chronic, and the symptoms will depend upon the quantity and duration of exposure.

MEDICAL CONDITIONS AGGRAVATED BY EXPOSURE: Inhalation exposure will aggravate preexisting respiratory ailments. Skin contact may aggravate preexisting dermatitis. Prolonged exposure to vapors may aggravate preexisting liver dysfunction.

RECOMMENDATIONS TO PHYSICIANS: Treat symptoms. Determine blood lead levels and remove patient from work area if elevated above 50 μg/100 g blood.

12. ECOLOGICAL INFORMATION

ENVIRONMENTAL STABILITY: The direct environmental stability of this product has not been determined. The lead and chromium content will be persistent in the environment, even if the organic portion decomposes.

EFFECT OF MATERIAL ON PLANTS OR ANIMALS: The direct effect of this material on plants and animals ha not been determined. The lead and chromium content are toxic to plants and animals at low concentrations.

EFFECT OF CHEMICAL ON AQUATIC LIFE: The direct effect of this product on aquatic life has not been determined. The lead and chromium content are toxic to aquatic life at low concentrations.

13. DISPOSAL CONSIDERATIONS

PREPARING WASTES FOR DISPOSAL: Waste disposal must be in accordance with appropriate Federal, state, and local regulations. This chemical, if unaltered by the handling, may be disposed of by treatment at a permitted facility or as advised by your local hazardous waste regulatory authority.

EPA WASTE NUMBER: D001, D008 (Pb), and D007(Cr).

14. TRANSPORTATION INFORMATION

THIS MATERIAL IS HAZARDOUS AS DEFINED BY 49 CFR 172.101 BY THE U.S. DEPARTMENT OF TRANSPORTATION

PROPER SHIPPING NAME: Flammable liquids, poisonous, n.o.s. (Cyclohexanone, Methylene Diphenyl Isocyanate)

HAZARD CLASS NUMBER and DESCRIPTION: 3, Flammable Liquid

UN IDENTIFICATION NUMBER: UN1992

DOT LABEL(S) REQUIRED: Flammable Liquid, Poison

PACKAGING GROUP: I

EMERGENCY RESPONSE GUIDE NUMBER: 28

MARINE POLLUTANT: Yes

15. REGULATORY INFORMATION

SARA REPORTING REQUIREMENTS: THE CHEMICAL(S), LISTED IN THE HAZARDOUS INGREDIENTS SECTION (Section 2), WHICH ARE MARKED WITH AN (*) ARE SUBJECT TO THE REPORTING REQUIREMENTS OF SECTION 313 OF TITLE III OF THE SUPERFUND AMENDMENTS AND REAUTHORIZATION ACT.

TSCA INVENTORY STATUS: The chemicals in this compounds ARE listed on the TSCA Inventory. Unlisted chemicals are marked with an † in Section 2.

CALIFORNIA PROPOSITION 65: This product contains chemicals known to the State of California to cause cancer.

CERCLA REPORTABLE QUANTITIES (RQ):

CHEMICAL NAME	REPORTABLE QUANTITY (lbs)
Lead	1
Methylene Diphenyl Isocyanate	1
Cyclohexanone	5000
Trichloroethylene	100
Chlorobenzene	100

STATE REGULATORY INFORMATION: The following states have designated that specific chemicals in this product are covered under specific state regulations:

Alaska - Designated Toxic and Hazardous Substances: YES
California - Permissible Exposure Limits for Chemical Contaminants: YES
Florida - Substance List: YES
Illinois - Toxic Substance List: YES
Kansas - Section 302/313 List: YES
Massachusetts - Substance List: YES
Minnesota - List of Hazardous Substances: YES
Missouri - Employer Information/Toxic Substance List: YES
New Jersey - Right to Know Hazardous Substance List: YES
North Dakota - List of Hazardous Chemicals and Reportable Quantities: YES
Pennsylvania - Hazardous Substance List: YES
Rhode Island - Hazardous Substance List: YES
Texas - Hazardous Substance List: YES
West Virginia - Hazardous Substance List: YES
Wisconsin - Toxic and Hazardous Substances: YES

LABELING (Precautionary Statements):
SIGNAL WORD: **DANGER!**
TARGET ORGANS: **RESPIRATORY SYSTEM, KIDNEYS, LIVER, SKIN, BLOOD**
HAZARD: **Flammable. Harmful if swallowed, absorbed through skin, or inhaled. Carcinogenic. Keep away from heat, sparks, flame. Keep in tightly closed container. Use with adequate ventilation.**

ventilation.

16. OTHER INFORMATION

PREPARED BY:
CHEMICAL SAFETY ASSOCIATES, Inc.
9163 Chesapeake Drive,
San Diego, CA 92123-1002
619/565-0302 DATE OF PREPARATION: April 4, 1994

INDEX